濃情美味

~為最愛的人下廚

黃淑儀

FAVORITE FOOD
FOR THE BELOVED

Foreword
代序

有朋友對我說，如果有機會去加拿大，一定要拜訪Gigi。Gigi，我一定會探，但奇怪，我的朋友為何滿眼期待、暗暗吞嚥口水。細問之下，原來她一定要我試Gigi炮製的豬扒，非常非常美味！

那次我去不成加拿大，幸好我是有口福的，不久後Gigi回港拍劇。

有一天，Gigi邀請我到她的家吃頓「家常便飯」，犀利囉，這頓「家常便飯」有十二道菜式，Gigi還設計了一張menu，我們只有十個人，可想而知，當天我們是非常滿足、愜意。

開飯時，我看見有燜牛腱，心想：很普通啫，我經常都會食。誰知，嘗過一口後，不得了，少少辣，又香又滑，與餐館吃到的味道完全不同。究竟下了甚麼調味料？秘訣是蝦膏，我完全猜不到。這晚的鹹豬手亦好精彩，我不懂怎樣形容它怎樣美味，總之比餐館好味得多。當晚的甜品「焦糖蛋」亦是另一個驚喜。

我好幸運能嘗過Gigi炮製的菜式，所用的心思、食材配搭，每每令客人驚喜、讚嘆！

羅蘭

A friend once told me, if you have a chance to visit Canada , make sure you drop by at Gigi's. However, I was curious to find her expression one of longing, and she seemed to be salivating! I then discovered, she wanted me to taste Gigi's special but outstanding pork chops.

I didn't get to visit Canada on that occasion. I was nevertheless blessed, for shortly after, Gigi returned to HK for film work.

One day, Gigi invited me to her home for some "simple home cooked food", or so she said. It turned out to be a 12 - course meal. Gigi even designed a menu card even though there were only ten of us. Needless to say, we were very satisfied that day.

At dinner, I saw there was the dish, braised beef shin. I thought to myself, this is a common dish. But, when I tried it, it was a little spicy and extraordinarily smooth and aromatic, so different from what is served in restaurants. What seasoning was added to it? The secret was prawn paste. I could not have guessed it! The salted pig's trotter was vey tasty as well. I fail in words to describe how flavorful it was, it was on the whole much more succulent than what any restaurant could offer. The dessert for the evening, crème caramel, was another pleasant surprise.

I am very blessed to have savoured Gigi's cooking. Her choice of ingredients, and the thoughtfulness applied to her cooking were real treats to her guests.

Law Lan

Over 40 years ago, a refreshing face appeared on Rediffusion TV. In particular, the bucktooth made her look so young and innocent. The young girl made an indelible impression on me.

I was then searching for some new faces for a movie. I then tried to request to "borrow" her from Rediffusion TV, but alas, I was rejected.

Lo and behold, many decades later, I would meet Gigi in TVB. Even with the limited opportunities for us to work together, I had the chance to savor the delicacies she brought to share with her colleagues.

Gigi's culinary skills is legendary. The most endearing thing about her though is the kindness emitted from her heart.

I remember once when Gigi heard that I was in Vancouver, she prepared a stupendous banquet, inviting many good friends to celebrate my birthday for me, for which I was very moved.

A good actress and a superb culinary expert, she has published ten cookbooks, preserving guaranteed good food.

This time, for me to have a chance to write this foreword for her cookbook is another opportunity for our cooperation . It also demonstrates my admiration for her in years passed was not in vain!

Wu Fung

40 多年前，在麗的電視螢幕上出現一位清秀小姑娘，尤其是她那隻凸出來的兔仔牙，特顯稚氣，給我留下深刻印象！

當時，我正為新戲找新面孔，於是向麗的電視借人，可惜被拒⋯⋯

沒想到，幾拾年後，會與芝芝在 TVB 重逢，雖然我們合作的機會不多，但仍然有機會嘗到她帶回公司與同事分享的美食！

芝芝的廚藝了得，眾所周知，但她最令人窩心的，是她發自內心的善意。

記得有一年，芝芝知道我在溫哥華，特意在家裏準備滿桌佳餚，請了一班老朋友為我慶祝生日，令我感動難忘！

芝芝除了是一位好演員，更是一位烹飪專家，烹飪書出到第十本了，朋友都説，只要跟着她的烹飪書去做，一定可以煮出似模似樣的餸菜！

今次為她的新書寫序，總算是另一種合作方式，也證明我當年對她的欣賞，眼光不錯！

胡楓

已經第十本了！

每一本，我都用心去做，每一道菜，我都要求份量要準確，製作過程要簡化！希望大家能一看就明，一試就行！

無可否認，這當中，仍有很多改善的空間，我虛心的向大家討教，讓我有進步的機會！

近年拍攝電視劇的工作也頗繁忙，尤其是拍畢古裝宮闈劇「萬凰之王」，身心疲累！只想好好休息，不接任何工作！

拍畢「On call 36 小時」經歷劇中人喪失親人之痛！久久不能自己！

我想起父親！

一個不苟言笑，不會與家人表達愛意的爸爸！在我剛學識煎魚時，賞我五毫子（五毛錢）！令我從此對烹飪產生興趣！改變我一生！

他嘗到我做的饅頭，竟然會説：這麼棒的饅頭，外面是絕對買不到的！那陣子，我沒有嫌做饅頭的工序有多繁複，一次又一次的做給爸爸吃！直到他離世！之後的十多年，我都沒有再做饅頭了！

今日，做饅頭的方法改良了，但口感依然，為了紀念逝世十年的父親，我積極籌備第十本烹飪書，加入部分爸爸愛吃的菜，希望他喜歡！也希望你們都喜歡！

黃淑儀

As with each of my cookbook, I kept the quantity of ingredients and cooking method of every dish precise to ensure that the recipes are easy to follow.

However, acknowledging that there is certainly room for improvement, I humbly ask for feedback and advice from all of you.

The filming work for TV dramas have undoubtedly taken up much of my time and energy. It was particularly so during the filming of "The Curse of the Royal Harem". It so exhausted me that I wished I could rest from all work!

After filming "On Call 36 hours", I was emotionally drained having played the role of someone who experienced the loss of loved ones. It took a while to regain my balance and composure.

It was then that memories of my late father sprang to mind – someone who was solemn and hardly knew how to communicate his love with his family.

I recalled when I started to learn how to pan-fry fish, he rewarded me with 50 cents, which roused my interest in cooking. As a result, my life changed.

He savored the buns I made and unexpectedly commented, "there is no way you could buy these flavorful buns anywhere!" For that period, it didn't matter how tedious making the buns were, again and again I made them for him, until he passed away.

For the next ten odd years, I did not make buns again.

Today, though the method of making buns is refined, the original texture remains. For my father's 10th death anniversary, I put my heart and soul into publishing this 10th Cookbook, adding some of his favorite dishes. May all of you savor them as he did.

Gigi Wong

目錄
CONTENTS

爸爸愛吃的菜

My Father's Favorite Dishes

爸爸，激勵了我對烹飪的興趣。

由我九歲時煎了一條又香又靚的紅衫魚，爸爸賞了我五毫子開始，到我學懂了蒸饅頭，爸爸嘗過後竟然讚道：這麼棒的饅頭，在外面絕對買不到。即使當時蒸饅頭的步驟繁複，我亦樂此不疲，經常蒸饅頭給爸爸享用。時光荏苒，我對烹飪的興趣有增無減，除了愛煮外，也因為有愛吃的知音人。

爸爸除饅頭外，也愛吃一些家鄉口味的小菜和點心，如蝦醬蒸豆腐、茼蒿蝦皮粉絲煲、紅豆鬆糕等等，希望你們也喜歡。

It was my father who aroused my interest in cooking.

It began with my father rewarding me with 50 cents for having fried a red snapper (at 9 years old) to perfection. When I learned to make steamed buns, he complimented me, " it is impossible to buy such good buns elsewhere!" Notwithstanding the complicated steps involved in making buns, I didn't grow tired of making them for my father. Till today, my interest in cooking has not waned and it's not for interest's sake, but it's also for people who appreciate food.

Besides liking steamed buns, my father loved to savor some appetizers and snacks from our hometown, such as Steamed Beancurd with Shrimp Paste, Braised Crown Daisy and Dried Tiny Shrimps with Mungbean Vermicelli and Steamed Red Bean Cake. I hope you will like them too.

Steamed Beancurd with Shrimp Paste

蝦醬蒸豆腐

傳統的蒸肉餅，加了醃料，加上配料就行了，但我在肉餅下墊了豆腐，令肉質更鬆軟；而豆腐沾到肉味，亦更豐富了它的味道層次。

Adding beancurd underneath the traditional steamed meat patty not only tenderizes the meat, but it also absorbs the essence of the meat to make it more flavorful.

材　　料

豆腐	2 塊（切粗粒）
免治豬肉	112 克
蝦醬	2 茶匙
蒜茸	2 湯匙
薑茸	1 湯匙
紅椒	1 隻（切粒）
葱花	1 湯匙
熟油	適量

Ingredients

2 pieces beancurd (cubed)
112 g minced pork
2 tsps shrimp paste
2 tbsps chopped garlic
1 tbsp chopped ginger
1 red chili (diced)
1 tbsp chopped spring onion
cooked oil

醃　　料

糖	2 茶匙
生抽	2 茶匙
粟粉	1 湯匙
麻油	適量
水	3 湯匙

Marinade

2 tsps sugar
2 tsps light soy sauce
1 tbsp cornflour
sesame oil
3 tbsps water

做　法

1. 豆腐粗粒鋪在碟底。
2. 免治豬肉加醃料後，將蝦醬、薑茸、蒜茸和紅椒粒拌進肉內，再鋪在豆腐上。
3. 用微波爐保鮮紙包好，在保鮮紙面刺幾個孔，放進微波爐內「叮」4 分鐘。
4. 灑上葱花，淋上熟油，即可享用。

Method

1. Lay the cubed beancurd on the bottom of a plate.
2. Combine the marinated minced pork with shrimp paste, chopped ginger, chopped garlic and diced red chilies. Spread it over the beancurd.
3. Wrap it with the microwave cellophane film. Poke a few holes in the film. Place it in a microwave and cook for 4 minutes.
4. Sprinkle with the spring onion with hot oil to serve.

冰豆腐 frozen beancurd

✻ Remarks ✻

· 這道菜可隔水蒸 10 分鐘，同樣色香味美。

· 豆腐買多了怎麼辦？留待第二天又會不新鮮。不妨將整磚豆腐放進冰格，吃時才解凍、切塊。冰凍過後的豆腐會呈現一層層，烹調時會如海綿般吸滿汁液，宜用於火鍋、燜、燴等。

· This dish is just as delicious if it is steamed over water for 10 minutes.

· Freeze the extra beancurd blocks. Thaw them only when cooking. They will have a layered texture. Cut into pieces for cooking and they will soak up the essence of the sauce and soup. Great for using it in hot pot and braising.

Braised Crown Daisy and Dried Tiny Shrimps with Mungbean Vermicelli

茼 蒿 蝦 皮 粉 絲 煲

這個家鄉菜，清簡、惹味，是盛夏的開胃小菜。
A simple and tasty family dish as a superb summer appetizer.

材 料

茼蒿	300 克（洗淨）
蝦皮	75 克（用薑片、葱度加過面水浸兩小時，以去腥鹹味）
五花腩	150 克（切絲，以胡椒粉、鹽少許略醃）
粉絲	1 紮（剪成兩截）
水	1 杯

Ingredients

300 g crown daisy (washed)
75 g dried tiny shrimps (soaked in water with sliced ginger and sectioned spring onion for 2 hours to get rid of the fishy smell and saltiness)
150 g pork belly (shredded; marinated with pepper and salt briefly)
1 bunch mungbean vermicelli (cut into halves)
1 cup water

調味料

鹽	半茶匙
糖	半茶匙
蠔油	1 湯匙
老抽	1 湯匙
胡椒粉	少許

Seasoning

1/2 tsp salt
1/2 tsp sugar
1 tbsp oyster sauce
1 tbsp dark soy sauce
pepper

做　法

1. 用白鑊以中慢火烘蝦皮至微黃，有香味，兜起備用。

2. 煲內放 2 吋水，加入 1 湯匙油和 1 茶匙鹽，下茼蒿，灼約 1.5 分鐘，盛起。

3. 用 1 湯匙油爆香五花腩，中火兜散，加入蝦皮，炒勻，加水，改用砂煲或鐵鍋盛起，轉大火，煮沸後，加調味料和粉絲，至粉絲軟身後，最後加入茼蒿，水分快乾前熄火。

Method

1. Stir fry the dried tiny shrimps without oil over medium low heat until fragrant. Remove and reserve for use.

2. Put water, about 2 inch high, in a pot. Add 1 tbsp of oil and 1 tsp of salt. Put in crown daisy and cook for 1 1/2 minutes. Take out.

3. Fry the pork belly in 1 tbsp of oil over medium heat till fragrant and loosened. Add the dried tiny shrimps. Stir well and add water. Turn them into a clay pot or iron pot. Turn up the flame to bring it to the boil. Add the seasoning and vermicelli. When the vermicelli is softened, add the crown daisy. Switch off the flame before the sauce is dried up.

Light & Easy Breakfast
Steamed Buns
with Soybean Milk

輕 怡 早 點
豆 漿 伴 饅 頭

自己做饅頭、豆漿，最有成就感，
想着營養，想着健康，你一定做得到！
如想做較細或較大份量的豆漿，
記着黃豆與水的比例是 1 杯已浸至發脹的黃豆配 2 杯水，
要懂得執生！
饅頭的材料及做法程序看似複雜，
但，只要按部就班，逐步逐步依照我的指示去做，
質感、味道都一流！
平日不苟言笑的父親，就曾經拿着我的饅頭說：
這是有錢都買不到的美味饅頭！

Great satisfaction comes from making buns and

soybean milk on your own. For the sake of health, we will do it!

To adjust the quantity of soybean milk produced,

just remember the soybean and water ratio is:

1 cup of soaked and enlarged soybeans to 2 cups water.

The ingredients and method used for making the buns may seem complicated;

however, as long as you follow the steps, the texture and flavor are excellent!

My late father, though solemn most times,

once commented on the buns I made,

"this is something not even money can buy."

材　料		Ingredients	
A		A	
熱水	4 湯匙	4 tbsps hot water	
糖	半茶匙	1/2 tsp sugar	
乾酵母	1 湯匙	1 tbsp dried yeast	
B		B	
油	2 湯匙	2 tbsps oil	
牛奶	1 杯	1 cup milk	
C		C	
麵粉	6 杯（篩勻）	6 cups flour (sifted)	
糖	3/4 杯	3/4 cup sugar	
蛋	2 個	2 eggs	

經過 5 分鐘後，
酵母已發脹一倍多。
After 5 minutes,
the yeast has risen double.

做 法

1. 預熱焗爐 100°C/200°F。
2. 將 A 料的糖放入熱水內，攪拌幾下，糖溶掉後，水溫已降成溫水，再把酵母加入，待 5 分鐘後發至一倍多。
3. 將 B 料混合，放入微波爐以高溫「叮」2 分鐘。
4. 將 C 料麵粉及糖混合拌勻，放在工作檯面，中間挖一個洞，蛋置中間，再趁熱把「2」、「3」加入其中，搓 10 分鐘。
5. 將粉糰分成小糰，做成喜愛的形狀，放在紙杯內，再排列在蒸盤上。
6. 焗爐熄火，放入蒸盤，讓其吸取餘溫，發酵 45 分鐘。
7. 燒開水，將「6」取出蒸 12 分鐘即可。

Method

1. Preheat the oven to 100°C/200°F.
2. Put sugar into the hot water in the Ingredients A combination. Stir to melt the sugar. Till the water temperature has dropped to warm, put the yeast in it. It will rise double in size after 5 minutes.
3. Mix Ingredients B and heat it in the microwave oven for 2 minutes over high heat.
4. Mix well the flour and sugar in Ingredients C. Place it on the countertop and make a hole in the middle. Pour "2" and "3" in it while it is hot, and knead for 10 minutes.
5. Divide it into small lumps and mould into any favorite shape. Put them in paper cups then place on a steaming tray.
6. Switch off the oven, place the steaming tray in it to absorb the residual heat to leaven for 45 minutes.
7. Boil water, remove "6" and steam for 12 minutes.

✻ Remarks ✻

- 這份量可做 20 個饅頭；餘下的饅頭可存放在冰箱，食時取出蒸軟即可！
- 不同品牌、氣候都會影響麵粉的乾濕度，建議不要一下子倒入所有材料 B，宜剩下少許，覺得不足才加入剩下的；如仍覺不足，可添加牛奶。
- The recipe can make 20 buns. The rest can be stored in the freezer for future steaming.
- However, as different brand and climate will affect the dryness and wetness of the flour, please do not use up all the ingredients in B at one go, spare a little to add it later if needed. If necessary, you may mix in a little more milk.

SoyBean Milk

豆漿

材　料

黃豆	300 克（浸過夜，洗淨）
水	約 10-12 杯
糖	3/4 杯（需要才加）
牛奶	1 杯

Ingredients

300 g soybeans
(soaked overnight; washed)
10-12 cups water
3/4 cup sugar (optional)
1 cup milk

一定要大力，才能搾出豆漿。
Use force to squeeze the milk through.

做　法

1. 黃豆 1 杯加水 2 杯放入攪拌機內，打爛，倒出豆漿；剩下的黃豆都以這個比例打爛（共做 4 次），以過濾布隔渣。
2. 豆漿加入糖，煮滾，熄火，加牛奶，即可飲用。

Method

1. Blend 1 cup of soybeans and 2 cups of water in a blender. Pour out the soybean milk; blend the rest of the soybeans in this proportion (repeat 4 times). Sift through a piece of maslin cloth.
2. Add sugar and bring to the boil. Switch off the flame and add milk to serve.

Mutton Soup with
Huai Shan, Qi Zi
and Ba Ji

淮 杞 巴 戟 羊 肉 湯

材　料

淮山	50 克	（略浸洗）
杞子	30 克	（沖洗乾淨）
巴戟	50 克	（略浸洗）
羊肉	1 斤	（斬件、汆水 20 分鐘）
龍眼肉	30 克	（沖洗乾淨）
薑	1 大塊	（拍扁）
水	16 杯	

Ingredients

50 g Huai Shan (washed briefly)

30 g Qi Zi (rinsed)

50 g Ba Ji (washed briefly)

600 g mutton
(chopped into chunks and scalded for 20 minutes)

30 g dried longan flesh (rinsed)

1 big chunk ginger (crushed)

16 cups water

做　法

1. 煮沸水，將全部材料倒入沸水內，猛火煲 10 分鐘，改用小火再煲 3 小時即可。

2. 隨個人口味下鹽調味。

Method

1. Boil the water. Put all the ingredients into boiling water. Boil over high heat for 10 minutes. Then turn to low heat and boil for 3 hours.

2. Season with salt accordingly.

✳ Remarks ✳

- 因為怕羊羶味，故很少以羊肉入饌，只是，這湯好處多多，而且易做，不妨一試！
- 此湯可滋陰、補腎、提神、養顏，男女均宜。
- Because of the pungent mutton smell, few people take to it. However, this soup has a lot of benefits. It is also simple to make, try it!
- This soup can aid the health of kidneys; enhance energy level and beautify skin. It is suitable for both men and women.

Steamed
Red Bean Cake
紅豆鬆糕

傳統的鬆糕勾起我不少美麗的回憶，
這是我父親喜愛的糕點。
添加紅豆在鬆糕裏，味道與材料都相
配，用日本的罐頭蜜紅豆可省卻煮豆
的工序，而且甜味剛剛好，不必再加
糖。
配上陳皮的橘子香，可以説是錦上添
花！

Traditional steamed red bean cake
brings back sweet memories to me as
I recall it was a favorite snack of my
father's.

Adding red beans in sponge cake blends
well with the flavor of other ingredients.
Using Japanese canned red beans saves
cooking time and its flavor is just right.
There is no need to add any sugar.
Additionally, using dried tangerine peel
provides extra aromatic enhancement!

材　料

A
麵粉	1 杯
粘米粉	半杯
發粉	1.5 茶匙
梳打粉	1 茶匙

B
溫水	1 杯
菜油	1 杯
蛋	8 個（打勻）
罐頭蜜紅豆	520 克
陳皮	半個（浸軟去瓤，剁茸）

Ingredients

A
1 cup flour
1/2 cup rice flour
1 1/2 tsps baking powder
1 tsp bicarbonate soda

B
1 cup lukewarm water
1 cup vegetable oil
8 eggs (beaten)
520 g canned sweet red beans
1/2 dried tangerine peel (softened
in water, pith removed and
chopped)

做　法

1. 將 A 料混合篩勻。
2. 將 B 料內的水、菜油及蛋，順序逐樣倒入「1」內，用打蛋器慢速拌勻。
3. 再加入蜜紅豆及陳皮茸，輕輕拌勻。
4. 將混合料倒入已抹油的糕盤內，猛火蒸 35 分鐘即成。

Method

1. Combine and sift Ingredients A well.
2. Pour the water, vegetable oil and eggs in Ingredients B in this order into "1". Mix well with egg beaters in low speed.
3. Add the red beans and dried tangerine peel, stir it gently.
4. Pour the mixture into a greased cake tin. Steam over high heat for 35 minutes. Serve.

為你帶來歡樂的菜式
Dishes which bring Joy!

不時會聽到：以食會友、民以食為天、辛苦揾來自在食等等。

在在説出，有甚麼事比吃更暢快、更愜意！

美食晃似鎂光燈般能發光發熱，能凝聚朋友和家人。每一次的聚會、每一天的飯菜，為所愛煮出美食，看他們吃得開心，你亦從中尋到樂趣。

Gigi 每一道菜都用心去做，能將複雜的菜式簡化但不失原味，務求令大家一看就明，一試就行！

Without a doubt, food generates much enjoyment!

Good food gathers family and friends around for fun and much satisfaction comes to the one who prepares them.

Gigi pours her heart into every dish. She transforms seemingly complex dishes into simple yet flavorful delicacies, so that the readers can understand them and succeed every time they use the recipes.

Porridge
明 火 白 粥

煲粥，任何白米均可。米與水的比例是 1：18；在這食譜我用兩杯米，所以要用 36 杯水。

煲粥方法各有不同，但煲粥令人卻步的原因可能是易「黐底」、容易煲燶。我的方法很簡單但保證粥綿、美味，只須在煲的過程保持粥「滾」就成。

若要加腐竹，需先用鹼水浸一會（鹼水與水的比例——鹼水 3 湯匙：水 4 杯），取出沖水，再煮成豆奶，待粥煮好後再加入。

做有味粥則加鹽 1 茶匙、糖 1 茶匙和少量麻油即可。

Use any type of white rice for making porridge. The ratio is: 1 part rice and 18 parts water. I use 2 cups of rice for this recipe so 36 cups of water is used.

One drawback for making porridge could be that it can easily coagulate at the bottom of the pots and become too thick. My method is simple yet it makes tasty and soft porridge: all you need to do is to ensure the porridge is always at boiling point during the cooking process.

If dried beancurd skin is added, first soak them in alkaline water (the ratio between alkaline water and water is 3 tbsps : 4 cups water). Remove them and rinse under water, then boil it to turn milky. Later, add to the cooked porridge.

Season with 1 tsp of salt, 1 tsp of sugar and some sesame oil to make savory porridge.

材　料

白米	2 杯
水	36 杯（9 公升）
油	1 湯匙
鹽	1 茶匙

Ingredients

2 cups white rice
36 cups (9 litres) water
1 tbsp oil
1 tsp salt

做　法

1. 白米洗淨後潷去水分，加油、鹽拌勻，再加過面水浸 2 小時。
2. 煮沸水後將米倒入，用杓不斷攪拌至水再滾起，不必冚蓋，可略收慢火，但必須保持「滾」的狀態；煲 1 小時即可。

Method

1. Wash the rice then discard the water. Add oil and salt to mix well. Cover it with water and soak for 2 hours.
2. Bring water to the boil, pour the rice in. Continuously stir it with a scoop until the water is re-boiled. Do not cover it but turn the heat lower. Maintain the boiling state however to boil for 1 hour.

The Boaters' Porridge
艇仔粥

材　料

白粥	1 鍋（做法及份量參閱 P.30 明火白粥）
免治牛肉	150 克（先用 1 湯匙梳打粉開水 2 湯匙醃一會，再調味、撻打幾十下）
江門排粉	1/8 包（用大火逐少炸至鬆起）
已浸發魷魚乾	1 條（用 4 杯水加 4 湯匙鹼水浸 2 日，啤水 1 小時，切粗絲）
砂爆豬皮	1 大塊（浸軟後以薑葱出水，切粗絲）
炸魚條	1 條（切粗絲）
葱花	1 杯
小粒花生	1 杯（慢火溫油炸 6 至 7 分鐘，臨起鑊前改大火炸 2 分鐘，迫出油分）

牛肉醃料

鹽	1 茶匙
糖	1 茶匙
胡椒粉	少許
水	4 湯匙

上味料

鹽	1 湯匙
糖	1 湯匙
水	4 杯

Ingredients

1 pot plain porridge (please refer to Page 30 on the method and ingredients used)

150 g minced beef (marinate with 1 tbsp of bicarbonate soda and 2 tbsps of water; season it and beat a dozen times)

1/8 packet Kong Moon rice sticks (fry over high heat in batches to loosen it)

1 treated dried squid (soaked in 4 cups of water and 4 tbsps of alkaline water for 2 days; rinse under running tap for 1 hour; then shredded thickly)

1 big piece deep fried pig's skin (softened in water then scalded with ginger and spring onions; shredded thickly)

1 piece deep fried fish fingers (shredded thickly)

1 cup chopped spring onions

1 cup small peanuts (fry over low heat in warm oil for 6 to 7 minutes; before scooping them up, turn up the heat to fry for 2 minutes)

Marinade for beef

1 tsp salt
1 tsp sugar
pepper
4 tbsps water

Seasoning

1 tbsp salt
1 tbsp sugar
4 cups water

已浸發魷魚，在凍肉舖有售。處理魷魚、豬皮，一點也不難，先用上味料煮兩分鐘，除令其有味外，還可將它們煮熟，吃時就更加安心。

You may buy the pre-soaked squid in frozen food stores. It is not difficult to treat the squid and pig's skin. Boil and cook them in seasoned water for two minutes to add flavor .

記住要買江門排粉，米粉才會炸得又鬆又脆；炸時米粉不宜多，待米粉炸至鬆起，撈出才放另一批。

Remember to buy Kong Moon rice sticks. The rice sticks after frying is puffy and crispy. Don't fry a large batch at one go. Scoop up the rice sticks only after it is puffy before the next batch.

免治牛肉裹上炸米粉，再按壓扁一點。

Coat the minced beef with rice sticks, then flatten slightly.

做　法

1. 牛肉醃過味後，與炸過的排粉揸成扁身牛丸。
2. 煮滾上味料，放入魷魚、豬皮，滾約兩分鐘即可取出。
3. 碗底放入牛丸、魷魚絲、豬皮絲和魚條絲，將白粥煮滾，倒入碗內，灑葱花和花生，再撒些胡椒粉即成。

Method

1. After marinating the beef, mix and mould with the rice sticks into some flatter beef balls.
2. Bring the seasoning to the boil, put the shredded squid and pig's skin in. Boil for 2 minutes. Set aside.
3. Place the beef balls, shredded squid, pig's skin and fish finger at the bottom of a bowl. Bring the plain porridge to the boil, pour into the bowl. Sprinkle the chopped spring onions and peanuts. Season with pepper.

Steamed Beancurd with Black Preserved Olive

欖菜蒸豆腐

白嫩的豆腐上，點綴了紅色的椒片、黑色的欖菜，再圍上翠綠的生菜，光是賣相，就夠吸引了！

何況，青菜豆腐，都能夠在平凡中顯示出極不平凡的營養呢！

Red chili slices, black preserved olive on whitish beancurd, with green lettuce surrounding it. This dish is attractively pretty!

Moreover, leafy vegetables and beancurd provides extraordinary nutrients from seemingly ordinary ingredients.

材 料

嫩豆腐	1 大塊
欖菜	2 湯匙
紅椒	1 隻（切片）
生菜膽	1 棵

Ingredients

1 big chunk beancurd
2 tbsps preserved olive
1 red chili (sliced)
1 stalk lettuce

做 法

1. 欖菜、紅椒片鋪在豆腐上，猛火蒸約 5 分鐘。
2. 煲內放 1 吋水，水內加油 1 湯匙、鹽 1 茶匙，放入菜膽，滾 1 分鐘即取出圍在豆腐旁。
3. 燒熱 2 湯匙熟油潷面即成。

Method

1. Arrange the preserved olive and red chili slices on top of the beancurd. Steam over high heat for 5 minutes.
2. Put 1 inch water in a pot. Add 1 tbsp of oil and 1 tsp of salt. Put the lettuce in and cook for 1 minute, take out and place around the beancurd.
3. Heat 2 tbsps of oil and pour on the beancurd. Serve.

Deep Fried Vegetarian Dumplings

炸素雲吞

有時，我會放縱自己，吃點脆卜卜的油炸品。雲吞餡料可切得精細或略剁，任隨尊便；但我這樣的做法、這樣的配料，不怕會炸得過老。只要你肯嘗試，炸功就是這樣練出來。因為大頭菜有鮮鹹味，故此不必加鹽加醋都很美味。

Sometimes, I indulge in some deep fried food. The wonton filling can be chopped briefly or sliced finely as you wish. My recipe however ensures that the accompaniments will not be over-fried. Try it to perfect your skill at deep frying. As the preserved vegetable is salty, it is no necessary to season it with salt or vinegar.

材　　料

廣東雲吞皮	適量
馬鈴薯	200克（放入適用於微波爐的碗內，加入過面熱水，「叮」15分鐘兩次，去皮，下調味料，再搓爛成茸）
冬菇	4朵（浸軟，切細粒）
紅蘿蔔	1/4條（去皮，切細粒）
大頭菜	1片（切細粒）
蛋白	1個

調味料

牛油	2湯匙
牛奶	2湯匙
鹽	

Ingredients

Cantonese dumpling skins (wonton skin)
200 g potatoes (Place them in a microwave bowl; add enough water to cover it and microwave for 15 minutes twice; remove the skin and season. Mashed.)
4 dried shittake mushrooms (softened in water and finely diced)
1/4 carrot (skin removed and finely diced)
1 piece black salted tuber (finely diced)
1 egg white

Seasoning

2 tbsps butter
2 tbsps milk
salt

將材料切粒、炒熟，雲吞皮包上餡料，完成。
Dice the ingredients and stir fry until cooked, wrap it in wonton skins.

41

做　法

1. 用 1 湯匙油爆炒冬菇、紅蘿蔔及大頭菜粒，與薯茸混合，攪勻。
2. 用雲吞皮包好餡料，以蛋白糊口。
3. 用小火油鑊，將雲吞炸至金黃色即可。

Method

1. Stir fry the dried mushrooms, carrot and black salted tuber till fragrant in 1 tbsp of oil, mix well with the mashed potato.
2. Wrap it in the wonton skin, seal the edge with the egg white.
3. Deep fry them over a low heat till they are golden brown.

先放一片雲吞皮試油溫；可以了，就逐粒逐粒雲吞放入油鑊，待雲吞皮金黃色就可以撈起，因餡料已預先炒熟。

Test the temperature of the oil by placing one slice of wonton skin in it. If it is acceptable, drop them individually into the wok, take them out when they turn golden brown.

Golden
Bitter Gourd
黃 金 涼 瓜

近年流行用鹹蛋黃製黃金蝦、黃金
蟹，我且把它換上涼瓜，引誘平時不
愛吃涼瓜的你！剩下的鹹蛋白可以用
來佐粥、或參考 P.59 鹹蛋菠菜的製
法。

Using salted eggs for dishes such as
Golden Prawns and Golden Crabs
has gathered momentum recently. I
have replaced it with bitter gourd in
order to tempt those who dislike the
vegetable. The left over salted egg
white is good to eat it with porridge,
or please refer to the recipe of Salted
Egg Spinach on page 59.

材　料
涼瓜　　　1 個（去瓤切片）
吉士粉　　2 湯匙
鹹蛋　　　4 個（隔水蒸熟，約 10 分鐘）
蒜茸　　　1 湯匙
紹興酒　　2 湯匙
葱花　　　1 湯匙
胡椒粉　　少許

Ingredients
1 bitter gourd (pith removed and sliced)
2 tbsps custard powder
4 salted eggs (steamed till cooked for
about 10 minutes)
1 tbsp chopped garlic
2 tbsps Shaoxing wine
1 tbsp chopped spring onion
pepper

做　法

1. 涼瓜沾上吉士粉，走油。
2. 分開鹹蛋白和蛋黃，趁熱用叉壓碎鹹蛋黃。
3. 用 2 湯匙油爆香蒜茸，加入蛋黃茸，炒至蛋黃起泡，加入涼瓜兜炒，灑胡椒粉，潷酒，最後灑上葱花即可。

Method

1. Coat the bitter gourd slices with custard powder and fry in oil.
2. Separate the salted egg whites and yolks, crush egg yolks with a fork while it is warm.
3. Fry the chopped garlic with 2 tbsps of oil till fragrant. Add the mashed egg yolks and fry till it bubbles. Stir in the bitter gourd slices; sprinkle with pepper and pour wine on the side of the wok. Sprinkle with the chopped spring onion to serve.

✳ Remarks ✳

宜買俗稱「雷公鑿」的苦瓜，它有苦瓜的獨特甘味；選購時以帶有光澤、表面沒有碰瘀、深綠色的為佳。

Select bitter gourd with a sheen and smooth body. Deep green ones are the best.

Stir Fried
Snow Peas with
Fresh Mushrooms

鮮菇炒雪豆

47

材　料

鮮冬菇　6朵（去蒂，用濕毛巾抹乾淨，
　　　　　切條）
雪豆　　225克（撕去老筋，汆水1分鐘）
薑米　　2湯匙
紹酒　　1湯匙
鹽　　　半茶匙

Ingredients

6 fresh mushrooms (stems removed;
wiped clean with a wring dry wet towel;
sliced)
225 g snow peas (strings removed;
blanched for 1 minute)
2 tbsps chopped ginger
1 tbsp Shaoxing wine
1/2 tsp salt

做　法

1. 用2湯匙油爆香薑米，加入雪豆及鮮
　 菇兜炒，灒酒，加鹽後快手兜勻即可
　 上碟。

Method

1. Fry the chopped ginger in 2 tbsps of
   oil till fragrant, add the snow peas
   and sliced fresh mushrooms to stir
   fry. Pour wine on the side of wok;
   season with salt then stir it briskly.
   Serve.

- 雪豆，在北美洲叫甜豆或蜜糖豆，真是名副其實的清爽甜美，彷如一個青春甜蜜的美女；它飽滿翠綠，所以炒時不能太熟，否則又黃又軟，失去甜豆爽脆的風采！
- 看似容易烹調的鮮菇炒雪豆，實際要精確掌握火候、時間，雪豆要炒至僅僅熟，才是顯功夫的真章！
- Snow peas are also known in North America as sweet peas or honey peas for its sweetness and crunchiness. Like a youthful beauty with her suppleness, snow peas are not to be over-fried, or they will lose the crunchiness while turning yellow and limp.
- It may look simple to fry fresh mushrooms with snow peas. However, to get the temperature of the flame and the required time right needs to be mastered.

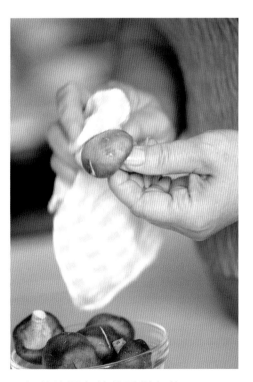

用扭乾的濕布抹乾淨鮮冬菇
Use a wring dry wet cloth to wipe clean the fresh mushrooms.

Eggplant in Special Sauce
醬汁拌茄子

茄子似海綿是很能吸收油分，宜用蒸或煮，吃時不油膩，不妨一試。茄子亦可以在飯滾後、快收乾水時直接放在飯面蒸熟，不過清除沾在上面的飯粒，要有耐性。涼拌茄子的醬汁亦可多樣化，例如只用浙醋、生抽、蒜茸、麻油、糖，又是別有一番風味！

為了美觀，在拍攝或宴客時我撕去茄子皮，其實茄子皮的食味也不賴！

Eggplant is like a sponge absorbs oil easily; hence it is best to either steam or boil it. Moreover, after the rice is boiled and as the water is being dried up, place eggplants on top of it to steam. Of course, clearing out the rice stuck onto the eggplants will take a lot of patience. Cold eggplant dish can have many varieties, such as using only the combination of Zhejiang vinegar, light soy sauce, chopped garlic, sesame oil and sugar is a flavor unto itself.

For aesthetic reasons, I tear off the skin of eggplants for photography or when I entertain my guests. In fact the taste of the skin is not bad at all.

材　料

茄子　　4 條（切成段）

醬　汁

芹菜　　1 棵（切碎）
葱　　　2 棵（切碎）
芫茜　　2 棵（切碎）
嫩薑　　4 片（切碎）
蘋果醋　2 湯匙
生抽　　4 湯匙
檸檬汁　1 湯匙
糖　　　1 茶匙
麻油　　1 茶匙
全部材料放入攪拌機拌勻成醬

Ingredients

4 eggplants (sectioned)

Sauce

1 stalk Chinese celery (chopped)
2 stalks spring onion (chopped)
2 stalks coriander (chopped)
4 slices young ginger (chopped)
2 tbsps apple cider vinegar
4 tbsps light soy sauce
1 tbsp lemon juice
1 tsp sugar
1 tsp sesame oil
Blend all the ingredients in a blender to make the sauce.

做　法

1. 茄子段放在碟上，隔水蒸 10 分鐘，去皮，再切成條。
2. 將醬汁淋在茄子上即成。

Method

1. Arrange the sectioned eggplants on a plate, steam over high heat for 10 minutes. Remove the skin and cut into strips.
2. Pour the sauce over the eggplants to serve.

Mixed Mushroom Salad

西式涼拌雜菌

如今我們真幸福，可以吃到多種新鮮菌類，不再是獨沽一味的冬菇乾貨了！不同的鮮菌有不同的香味、不同的口感，我們可以隨意視乎市場供應來變化這食譜的菌類用料，淋上汁液，簡單、別緻又美味！此菜亦可當作西餐伴碟用。

鮮菌的吸水力強，就像一塊海綿，所以千萬不能泡水，只能用扭乾的濕布抹乾淨，否則定會煮到水汪汪，注定失敗！

Nowadays, we are blessed with many kinds of fresh mushrooms instead of sticking to the dried mushrooms. Different fresh mushrooms possess distinctive flavors and textures. We can use any seasonal mushrooms for this dish. Pour sauces on it. It's as easy and delicious as that! It can also be used as an accompaniment in western dining.

Fresh mushroom is water absorbent, like a sponge. Hence, never soak them in water. Use a wring dry wet cloth to clean them, or it will be too watery!

材　料

蠔菇	225 克（用濕毛巾抹淨）
草菇	225 克（用濕毛巾抹淨、一開二）
靈芝菇	1 包（切除根部木屑）
洋葱	半個（切粗絲）
蒜頭	4 粒（切幼粒）
芫茜	2 棵（洗淨，切碎作裝飾用）

爆草菇料

薑	2 片
葱	2 棵（切度）
紹酒	2 湯匙

汁　料

芥末醬	1 湯匙
白酒醋	2 湯匙
糖漿	1 湯匙
橄欖油	4 湯匙
鹽、胡椒粉	各少許

做　法

1. 草菇處理法：爆香薑、葱，灒紹酒，倒入草菇，兜炒一會，盛起，棄去薑葱和汁液。
2. 先將汁料調好。
3. 用 1 湯匙油爆香蒜粒及洋葱，將所有菌類入鑊，炒至剛熟，倒入汁料兜勻，灑上芫茜即可享用。

Ingredients

225 g oyster mushrooms (wiped clean with wring dry wet cloth)

225 g straw mushrooms (wiped clean with wring dry wet cloth, halved)

1 pack shimeji mushrooms (stems cut off)

1/2 onion (shredded)

4 cloves garlic (chopped)

2 stalks coriander (washed and chopped for decoration use)

Ingredients for

Frying Straw Mushrooms

2 slices ginger

2 stalks spring onion (sectioned)

2 tbsps Shaoxing wine

Sauce

1 tbsp mustard paste

2 tbsps white wine vinegar

1 tbsp syrup

4 tbsps olive oil

salt and pepper

Method

1. Treating the straw mushrooms: Fry the ginger and spring onion till fragrant. Pour Shaoxing wine at the side of wok, stir for a while. Scoop up. Discard the ginger, spring onion and the sauce.

2. Make the sauce.

3. Fry the chopped garlic and onion in 1 tbsp of oil till fragrant. Add all the mushrooms in the wok. Stir fry till they are just cooked. Pour the sauce in and mix well. Sprinkle with the coriander to serve.

經處理後的草菇會去除霉味，突顯草菇鮮味。
The fresh taste of straw mushroom will be enhanced after treating it. The musty odor is removed.

菠菜兩吃

菠菜的可塑性非常高，用來煮、炒、涼拌，甚至批餅餡料均可。有時想嘗嘗新口味，可炮製這鄉土味濃的鹹蛋菠菜或帶點日式口味的芝麻菠菜卷。記着焓菜只用 2 吋水就可以，不用煲一大鍋水，只會徒浪費燃料。

Spinach Served Two Ways

Spinach is versatile. It can be boiled or stir fried and use as a cold salad or pastry filling. For a different take, try this earthy dish Salted Egg Spinach, or the Japanese flavored Sesame Spinach Wraps.

Only 2 inches of water is needed to blanch spinach, its not necessary to use a big pot of water.

Sesame Spinach Wraps
芝麻菠菜卷

材　料
菠菜　　　　　600 克（洗淨）
黑白芝麻　　　各 75 克（熱水沖洗乾淨，瀝
　　　　　　　去水分）
日式芝麻醬　　適量

道　具
壽司蓆　　　　一張

Ingredients

600 g spinach (washed)
75 g each of black and white sesame seeds
(rinsed in hot water; drained)
Japanese sesame sauce

Tool

1 piece of sushi mat

做　法

1. 煲內注入 2 吋水，加入油 1 湯匙、鹽 1
   茶匙後即放入菠菜，灼 1.5 分鐘即可取
   出，待冷。
2. 黑白芝麻分別放在白鑊中烘至微黃，倒
   出待用。
3. 菠菜放在壽司蓆中，捲實，切成喜愛形
   狀，沾上芝麻，放入已倒芝麻醬的碟內
   即可。

Method

1. Fill a pot with 2 inch high of water, add
   1 tbsp of oil and 1 tsp of salt to blanch
   the spinach for 1 1/2 minutes. Remove
   and set aside to let cool.
2. Stir fry the black and white sesame
   seeds in a dry wok till they turn slightly
   yellowish.
3. Place the spinach on a sushi mat, roll it
   up tightly. Cut into any shape, dip them
   in the sesame seeds and place them on a
   plate with sesame sauce to serve.

Salted Egg Spinach
鹹蛋菠菜

材　料

菠菜　600 克（洗淨後，用手摘斷，一開二）

鹹蛋　2 個（鹹蛋放入凍水內，滾 3 分鐘，熄火，焗 10 分鐘，去殼，切粒）

Ingredients

600 g spinach (washed and torn with hands into two)

2 salted eggs (place in cold water and boil for 3 minutes; switch off the flame and cover for 10 minutes; shells removed and diced)

做　法

1. 煲內注入 2 吋水，煮沸後加入 1 湯匙油、1 茶匙鹽，放入菠菜，冚蓋，煲 1.5 分鐘即取出，瀝乾水分。

2. 把鹹蛋粒灑在菠菜上即可享用。

Method

1. Fill a pot with 2 inch high of water, add 1 tbsp of oil and 1 tsp of salt to blanch the spinach for 1 1/2 minutes. Remove and drain.

2. Sprinkle the diced salted eggs on the spinach to serve.

✳ Remarks ✳

- 不要摘斷菠菜，切菜菜卷時才不會散開。
- Don't break the spinach or they will spread out when cut into rolls.

Braised Prawns
with Mungbean Vermicelli
粉絲蝦煲

材　料

中蝦	450克（剪去蝦頭尖刺、蝦鬚，挑腸）
粉絲	2 小紮（浸軟，剪斷）
蒜頭	2 粒（切片）
薑	4 片
葱	2 棵（切度）
中芹	1 棵（洗淨切度）
酒	2 湯匙

調味料

蠔油	1 湯匙
生抽	2 茶匙
糖	1 茶匙
黑椒粉	1 湯匙
水	1 杯

Ingredients

450g medium-sized prawns (prawns' heads and whiskers removed; deveined)
2 batches mungbean vermicelli (softened in water and cut)
2 cloves garlic (sliced)
4 slices ginger
2 stalks spring onion (sectioned)
1 stalk Chinese celery (washed and sectioned)
2 tbsps wine

Seasoning

1 tbsp oyster sauce
2 tsps light soy sauce
1 tsp sugar
1 tbsp black pepper
1 cup water

我煮這菜式粉絲的比例是多了一點，因為粉絲吸收了蝦的鮮味，往往客人都好快把粉絲掃光，故此索性就下多點吧！
記住粉絲要煮得煙韌，蝦肉要爽口，是這小煲的特點！

I usually add more mungbean vermicelli as it soaks up the sweetness of prawns. It is a popular dish amongst my guests. Just remember to prepare mungbean vermicelli to be al dente and the prawns firm.

記得要將粉絲浸軟後剪斷，否則吃時會有點麻煩。

Remember to soak and soften mungbean vermicelli first before cutting it, or it is messy when serving.

做　法

1. 用略多油將蝦以半煎炸方法走油，每邊約 30 秒，取出待用。
2. 取少許泡過蝦的油，爆香蒜片、薑片和葱度，放入蝦，瓚酒，兜勻蝦及料頭後加入調味料，煮沸後放入粉絲，冚蓋以小火煮 5 分鐘，最後加入中芹，即可上枱或盛入瓦煲內奉客。

Method

1. Pan fry the prawns in oil for about 30 seconds on each side. Reserve for use.
2. Fry the garlic, ginger and spring onion in a little oil used for cooking the prawns till fragrant. Put the prawns in. Sprinkle the wine on the side of the wok. Stir the prawns and aromatics well. Add the seasoning. Bring it to the boil then place the vermicelli in. Cover and cook over low heat for 5 minutes. Lastly add the Chinese celery to serve.

Baked Prawns in Butter and Minced Garlic

 牛油蒜茸焗大蝦

蝦，雖然是普通的材料，但用點心思和時間，就可做出這款食得又睇得的焗大蝦了。這菜式看似難做，但只需將蝦略煎，然後放入焗爐，在預定時間就有得吃了，非常簡單，大家來試試吧！

With a little effort seemingly ordinary prawns can be turned into a great dish. It may look difficult to prepare at first glance, but pan fry the prawn briefly, then place it in an oven, within a specified time, it is done!

材　　料

大蝦　　　　8 隻（劏開背部成雙飛狀，去腸）

釀　　料

牛油溶液	2 湯匙
蛋黃	1 個
白麵包糠	5 湯匙
混合香料	1 湯匙
炒香蒜茸	2 湯匙
炒香乾葱	2 湯匙
鹽	1 茶匙
胡椒粉	少許
辣椒粉	少許
	（視乎口味加減）

Ingredients

8 big prawns (made a deep incision along the back of the prawn, cutting virtually halfway through the meat to create a butterfly shape and deveined)

Stuffing

2 tbsps melted butter

1 egg yolk

5 tbsps white breadcrumbs

1 tbsp mixed herbs

2 tbsps fried chopped garlic

2 tbsps fried dried shallot

1 tsp salt

pepper

chili powder (on personal taste)

挑蝦腸、剪開蝦背，再填釀料其實好簡單。挑蝦腸，用一枝牙籤就搞定。

It is actually vey simple to devein prawns. Open the back. Use a toothpick to devein prawns. Then spread with the stuffing.

做　法

1. 預熱焗爐 160°C/320°F。
2. 把釀料混合成糊狀，在蝦背塗上粟粉，再填釀料。
3. 用牛油起鑊，用中火把蝦背略煎，取出。
4. 放入焗爐內焗 10 分鐘即可。

Method

1. Preheat the oven to 160°C/320°F.
2. Mix the stuffing ingredients into a paste, stuff it on the back of prawns.
3. Heat the butter in a wok, pan fry the back of prawns over medium heat. Remove.
4. Place in the oven to bake for 10 minutes to serve.

Grilled Fresh Scallops

燒鮮帶子

烹調海鮮用最簡單的方法,配最基本的調味料,才能發揮其鮮味。

To have the fresh taste of seafood, it is best to use the simplest of seasoning and cooking method.

材　料
大急凍帶子	12 隻
番茜碎	1 湯匙

醃　料
鹽	半茶匙
胡椒粉	少許
檸汁	2 湯匙

醬　汁
白酒醋	2 湯匙
芥末醬	半茶匙
糖	1 茶匙
橄欖油	2 湯匙
鹽	1/4 茶匙
胡椒粉	少許

* 搖勻

Ingredients

12 defrosted scallops
1 tbsp chopped parsley

Marinade

1/2 tsp salt
pepper
2 tbsps lemon juice

Sauce

2 tbsps white wine vinegar
1/2 tsp mustard paste
1 tsp sugar
2 tbsps olive oil
1/4 tsp salt
pepper
*shake the above well

做　法

1. 帶子用醃料醃半小時,煎前索乾汁液。
2. 燒熱 2 湯匙油,收中火,把帶子兩面煎約 30 秒至 1 分鐘,加入醬汁,收慢火,輕手拌勻,讓每粒帶子都入味,最後灑上番茜碎即成。

Method

1. Marinate the scallops for 1/2 hour. Pat them dry with kitchen paper before frying.
2. Heat 2 tbsps of oil in wok, reduce to medium heat. Fry the scallops both sides. Each side for about 30 seconds to 1 minute. Add the sauce. Reduce the flame to low heat. Stir gently to allow the scallops to absorb the flavor. Sprinkle with the chopped parsley to serve.

Thai Style Fried Pomfret

泰式煎鯧魚

用泰式香料煎封鯧魚，惹味，能刺激胃口，
實在是下飯的好伙伴！
在北美洲通常只買到急凍的鯧魚，
幸而，近年來的急凍海產解冰後都能保持水準，
選擇就可以多一些了。只要挑到實肉的鯧魚，
加上調味，不難煮出美味。

Marinating a pomfret with Thai spices is both flavorful
and appetizing! It is wonderful to eat it with rice.
We only get frozen pomfret in North America.
However, its quality is good.
Choose fishes with firm flesh.
With some seasoning, it will turn out a delectable dish.

鯧魚　　　　　1 條約 900 克（劏洗乾淨，用少許鹽抹勻魚肚，在魚背剠三刀）

配　　料

蒜頭　　　　　4 粒（剁茸）
乾葱頭　　　　4 粒（切碎）
紅辣椒　　　　2 隻（切碎）
九層塔葉　　　1 棵份量（切碎）

汁　　料

魚露　　　　　1 湯匙
生抽　　　　　1 湯匙
米酒　　　　　1 湯匙
紅咖喱醬　　　1 湯匙
青檸汁　　　　1 個份量
糖　　　　　　1.5 湯匙
水　　　　　　半杯

魚要烹調得美味，實在需要一些技巧：在魚肚內抹鹽，有辟腥的作用；在魚背上剠三刀可令魚快點熟。

Some techniques are required to make a delicious fish dish: rub salt in the belly of fish to get rid of the fishy odor; making 3 cuts at the back of fish speeds up the cooking process.

Ingredients

1 pomfret (900 g) (gut and washed; rub the belly of fish with salt, make 3 cuts at the back of fish)

Aromatics

4 cloves garlic (chopped)
4 shallot (chopped)
2 red chilies (chopped)
1 bunch Thai basil leaves (chopped)

Sauce

1 tbsp fish sauce
1 tbsp light soy sauce
1 tbsp rice wine
1 tbsp red curry paste
juice of 1 lime
1 1/2 tbsps sugar
1/2 cup water

榨汁前將青檸用力按壓，會更易榨出汁液。如沒有榨汁器，那就地取材，用叉代替吧！

Press lime hard before squeezing its juice, this is to make it easier to extract the juice. If there is no juice extractor, please use a fork instead.

做　法

1. 用略多油以半煎炸烹調法把魚煎至兩面金黃，每邊約 2 分鐘，取出待用。
2. 燒熱 1 湯匙油，改用慢火爆香配料，再倒入汁料，煮開淋在鯧魚上即可。

Method

1. Use a little more oil to pan fry the pomfret till both sides are golden. Each side takes about 2 minutes. Remove for use.

2. Heat 1 tbsp of oil, fry the aromatics over low heat till fragrant. Pour the seasoning in, bring it to the boil then pour onto the fish to serve.

Wrapped Salmon
釀三文魚

這菜式的靈感女神是 Lisa，她是酒店的飲食部經理，對食物要求高，認識深，除了識食，也識做。擔任這個職位，絕對稱職！暇時亦會撚番兩味，大部分是煮西餐，亦是我煮西餐的盲公竹。

三文魚是難纏的傢伙，如作刺身享用，軟滑，甘美；如作熟食，若處理不當會變得乾硬、粗糙。如果沒有把握煎到僅僅熟，既有汁液，亦有魚肉軟滑感，最好試用這個方法，包你會愛上它。

三文魚的皮下脂肪好豐富，保留魚皮，好讓它發揮滋潤的功效。

It was Lisa who inspired this dish! As a F & B Manager in a hotel and highly knowledgeable in food, Lisa is my guide for western cooking.

Salmon is tricky to handle. When used in sashimi, it is smooth and flavorful whereas any mishandling of it in cooking will make it coarse, hard and dry. If we are unsure of mastering the fish to attain the level of moist perfection, this recipe will perhaps be much easier to tackle. You'll love it!

Keep the fish skin as the rich fat underneath it can keep the fish moist.

材　料

三文魚	2 塊（連皮）

醃三文魚料

橄欖油	1 湯匙
鹽	1 茶匙
黑椒粉	適量
法式芥辣	半茶匙

包三文魚料

日式麵包糠	1 杯
檸檬皮茸	1 個份量
百里香	2 茶匙
橄欖油	3 湯匙

Ingredients

2 pieces salmon (with skin)

Marinade

1 tbsp olive oil
1 tsp salt
black pepper
1/2 tsp French mustard

Salmon Wrap

1 cup panko breadcrumbs
zest of 1 lemon
2 tsps thyme
3 tbsps olive oil

做　法

1. 預熱焗爐 175°C/350°F。
2. 三文魚先抹橄欖油，灑上鹽及黑椒粉，再抹上法式芥辣。
3. 將包三文魚料混合，裹在三文魚魚皮上，焗 8 至 10 分鐘即成；亦可用光波爐焗 8 分鐘即可。

Method

1. Preheat the oven to 175°C /350°F.
2. Rub the salmon with olive oil, sprinkle with salt and black pepper. Rub it with the French mustard.
3. Combine the ingredients for the salmon wrap to cover the skin of salmon. Bake for 8-10 minutes to serve. Or, bake in a multi-purpose halogen cooking pot for 8 minutes.

Butter Lobster

奶 油 龍 蝦

奶油龍蝦是 80 年代北美洲最流行的食譜，當時的龍蝦才 $2.99 一磅，用 whipping cream 烹調又鮮又滑。對 whipping cream 陌生的我，馬上被吸引着，更馬上學會煮這道菜。

也有食肆用龍蝦蟹合拼來做招牌菜，其中的巧妙就是用 whipping cream，有相得益彰的效果。

Butter Lobster was a popular dish in the 80s in North America. Lobster cost only $2.99 per pound then. Prepared with whipping cream, it was fresh and smooth. Henceforth I was attracted to whipping cream. I promptly learnt to cook this dish. There are restaurants which use a combination of lobster and crabs as their signature dish. The secret lies in using whipping cream.

材　料

龍蝦	1 隻約 1.3 千克（斬件）
粟粉	4 湯匙
牛油	2 湯匙
紅葱頭	4 粒（切絲）
鹽	1 茶匙
全脂牛奶	2 杯
椰子油	半茶匙

Ingredients

1 lobster (about 1.3 kg)(cut into pieces)
4 tbsps cornflour
2 tbsps butter
4 shallots (cut into shreds)
1 tsp salt
2 cups whole milk
1/2 tsp coconut oil

因龍蝦上了粟粉，回鑊兜炒時與牛奶融合，汁液會很濃厚，不
會水汪汪。可用忌廉代替牛奶，汁液會更滑更香。
Lobster rubbed with cornflour will blend well with milk when
stirring with it.The sauce will be thick and not watery.
Whipping cream can replace milk to make a smoother and more
aromatic sauce.

做　法

1. 灑適量粟粉在龍蝦上，放進油鑊內走油，撈起。
2. 用牛油爆香紅葱頭，下鹽，要不停兜炒至香氣四溢，
   倒入全脂奶和椰子油，將龍蝦回鑊，兜勻即可上碟。

Method

1. Rub the lobster with some cornflour. Scald in oil for a
   while. Remove.
2. Fry the shallots in butter till fragrant, add salt, stir
   constantly till it's fragrant. Pour the milk and coconut
   oil in, put the lobster back in the wok, stir and serve.

Stewed Black Pepper Chicken Drumsticks

黑椒雞髀

楊煥素是孩子們的鋼琴老師，亦是我的烹飪知己。這個雞髀既易做又惹味，是孩子們的至愛。

因為每家爐頭火力各有不同，若 15 分鐘後鍋內仍有許多汁，宜把雞髀挾起，用猛火煮至汁液濃稠後，再倒入雞髀撈勻；若不取起雞髀，雞髀會燜至軟爛，外形不美之餘，更沒有口感。

Sue Wan was my children's piano teacher and my cooking pal. This recipe for chicken drumstick is both simple and tasty – a favorite of my kids.

Each stove has different temperature levels. Hence if after 15 minutes there is a lot of sauce left, it is then best to remove the chicken drumsticks and cook the sauce over high heat till it thickens. After that, add the drumsticks to stir well. If the drumsticks are not removed, they will be too mushy.

材　料

雞下髀　　900 克

調味料

海鮮醬　　3 湯匙
生抽　　　2 湯匙
老抽　　　1 湯匙
黑椒碎　　1 湯匙
　　　　　（可隨個人喜好增減）
水　　　　1 杯
薑　　　　2 片

做　法

1. 混合所有調味料。
2. 將雞下髀及調味料倒入煲內，猛火煲約 15 分鐘至汁稠即可。

Method

1. Combine well all the seasoning.
2. Add the drumsticks and seasoning into a pot, boil over a high heat for 15 minutes till the sauce thickens.

Ingredients

900 g chicken drumsticks

Seasoning

3 tbsps Hoi-sin sauce
2 tbsps light soy sauce
1 tbsp dark soy sauce
1 tbsp crushed black peppercorns
(as per personal taste)
1 cup water
2 slices ginger

Korean Style Chicken

韓風雞片

材　　料

雞柳	4 個（切片）
青、紅、黃燈籠椒	各半個（切塊）

醃　　料

蒜茸	1 湯匙
洋蔥泥	1 湯匙
生抽	1 湯匙
韓式醬料	2 湯匙
糖	少許
麻油	1 茶匙

Ingredients

4 chicken fillets (sliced)
half of each green, red and yellow
capsicum (cut into wedges)

Marinade

1 tbsp minced garlic
1 tbsp mashed onion
1 tbsp light soy sauce
2 tbsps Korean style sauce
sugar
1 tsp sesame oil

做　　法

1. 雞肉先用醃料醃起碼半小時。
2. 用 2 湯匙油爆香雞肉，待雞肉變色，倒入三色椒，快手兜勻，即可盛起。

Method

1. Marinate the chicken for at least 30 minutes.
2. Stir fry the chicken with 2 tbsps of oil till fragrant. When the color of the flesh changes, put in the capsicums. Stir briskly. Remove.

✳ Remarks ✳

- 三色椒炒至半生熟即可，入口才爽脆。
- 韓式醬料在一般超市都可以買到，較一般醃料味濃，故只需加少許糖中和一下即可，簡單易做。
- Stir fry the capsicums only till they are half cooked to keep the crunchiness.
- Korean style sauce can be found in most supermarkets. As its flavor is heavier than most marinades, add a little sugar to enhance the flavor.

Smoked Chicken
燻雞

燻雞可以熱食或凍食，為蒸雞加添一點不尋常的香氣。除了宴客，野餐時帶備手撕燻雞伴方包，令你的旅途變得不平凡；或可加入即食麵內，加添變化。

Eat it cold or hot, smoked chicken has a distinct taste when compared to steamed chicken. Besides serving it at parties, bring the dish along on picnics to go with some bread. It can also be added to instant noodles.

材　料
光雞　　　1 隻（約 900 克）
粗鹽　　　1.5 湯匙

燻雞料
米　　　　半杯
黃糖　　　2 湯匙
紅茶包　　2 個（或紅茶葉 1 湯匙）

Ingredients
1 chicken (about 900 g)
1 1/2 tbsps coarse salt

Smoked Chicken Ingredients
1/2 cup rice
2 tbsps brown sugar
2 black tea bags
(or 1 tbsp black tea leaves)

做　法

1. 將粗鹽抹勻雞內外，醃過夜，蒸前沖去鹽分，猛火蒸 20 分鐘。
2. 用錫紙墊鑊底，放入燻雞料，上放蒸架，把雞置蒸架上，開猛火，冚蓋燻 30 分鐘即可。

Method

1. Rub the body and cavity of the chicken with the coarse salt. Marinate it overnight. Before steaming, wash the salt off. Steam over high heat for 20 minutes.
2. Lay the bottom of a wok with aluminum foil. Place the smoked ingredients in. Put a rack above it for the chicken. Cover and smoke the chicken over high heat for 30 minutes.

✳ Remarks ✳

燻雞可涼拌、手撕，亦可熱食，絕對是宴客的佳餚。

Tear it with hands, smoked chicken can be eaten cold. It is good eaten hot as well - a good dish for entertaining guests.

因燻煙食物會造成濃煙，故在鑊底、鑊蓋內都要墊上錫紙，包好鑊蓋，完成後，只需將錫紙取走即可。

Smoking food emits thick smoke. Hence, cover the wok and lid with aluminum foil. When it is done, just remove the foil.

Steamed Chicken with Sand Ginger Sauce

蒸沙薑雞

我喜歡蒸整隻雞，較原汁原味；
蒸後的雞汁更可煮雞油飯，非常美味。
I love to steam a whole chicken
as it retains its original flavor.
Plus, its juice after cooking can also be used
for preparing Chicken Oil Rice.
Very tasty!

材　　料

雞	1 隻（重約 1.2 千克）
葱	2 條
薑	2 片

醃　　料

鹽	2 茶匙
沙薑粉	2 茶匙
雞粉	1 茶匙

Ingredients

1 chicken (about 1.2 kg weight)
2 stalks spring onion
2 slices ginger

Marinade

2 tsps salt
2 tsps sand ginger powder
1 tsp chicken powder

做　　法

1. 用醃料將雞內外抹勻，肚內塞入薑葱，最好醃過夜。
2. 隔水猛火蒸 20 分鐘。
3. 取去肚內薑葱，待涼，斬件上碟。

Method

1. Rub the marinade on the cavity and the outside of the chicken evenly. Stuff the ginger and spring onion. It is best to leave the marinade overnight.
2. Steam over high heat for 20 minutes.
3. Remove the ginger and spring onion. Set aside to let cool. Chop up the chicken and serve.

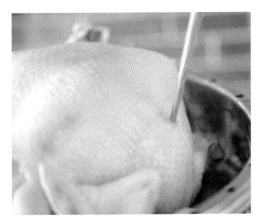

要測試雞是否已熟？可將筷子插入最厚肉的部位，如沒有血水流出，則代表雞已熟。

To test the doneness of chicken, just insert a chopstick into the thickest meat, if no blood is oozing out, the chicken is cooked.

Baked Pork Ribs
焗豬肋骨

處理豬扒是我的弱項，無論煎、炸、焗都只有肉味，但沒有質感—好似柴皮一樣粗韌。如今有「奇異果茸」的招數，肉質得以改善，當堂信心大增。因為奇異果內含有豐富的蛋白質溶解酵素 actinidin，用它來醃肉可令肉質變得柔軟。

I used to feel helpless handling pork chops. It was difficult to achieve overall tenderness. But with the addition of kiwi fruit, the texture is much improved. Kiwi fruits possess rich protein dissolving enzymes called actinidin. This works in softening the meat in marinade.

材　　料

豬肋骨　　　1 排

奇異果茸

奇異果　　　1 個（打爛）

調味料

甜椒粉　　　1 湯匙
混合香料　　1 湯匙
鹽　　　　　半茶匙
黑椒碎　　　1 湯匙
紅糖　　　　2 湯匙
BBQ 醬　　　適量

Ingredients

1 pork ribs

Minced Kiwi Fruit

1 kiwi fruit (smashed)

Seasoning

1 tbsp paprika
1 tbsp mixed herbs
1/2 tsp salt
1 tsp crushed black peppercorns
2 tbsps brown sugar
BBQ sauce

做　法

1. 預熱焗爐 175°C/350°F。
2. 用奇異果茸醃豬肋骨半天。
3. 豬肋骨抹上甜椒粉、混合香料，再下鹽、黑椒碎、紅糖醃兩小時。
4. 放入焗爐焗兩小時後再搽上 BBQ 醬，改用上火，燒至微焦即可。

Method

1. Preheat the oven to 175°C/350°F.
2. Marinate the pork ribs with the mashed kiwi fruit for half a day.
3. Rub the ribs with paprika and the mixed herbs; add salt, crushed black peppercorns and brown sugar to marinate for 2 hours.
4. Bake in the oven for 2 hours. Spread the BBQ sauce on it, turn up the heat to grill till they are slightly burnt.

* Remarks *

· 如用光波爐，只須焗45分鐘，省下不少時間。
· 很奇妙，用奇異果茸作鬆肉劑，肉不但沒有沾上奇異果味，還會更加鬆軟。
· Bake for only 45 minutes in a multi-purpose halogen cooking pot.
· Kiwi fruit is amazing. As a meat tenderizer, it doesn't stain the flavor of meat but instead will tenderize it.

Lisa's Pork Chop
Lisa 豬扒

對，煎豬扒的確有點難度，但嘗過 Lisa
為我煎焗的豬扒，覺得一定要向大家介
紹，好讓你能吃到原汁原味的特式豬扒！
焗後的青蘋果一點也不酸，還有肉汁香
哩。

喜歡的話，可以芥末伴食！

Pan-frying pork chops can have its fair
share of difficulties. However, I would like
to recommend to you Lisa's Pork Chop. It
retains the juice and flavor as well as laced
with the aroma of the meat. The green
apples are not sour. Serve this dish with
mustard, if you like.

材　料

厚豬扒連骨	4 塊
	（用鹽、黑椒碎略醃）
紅葱頭	1 粒（切幼粒）
百里香	20 條
青蘋果	3 個
	（去皮去芯，一開八）
白酒	半杯

伴　碟

煙肉	4 條（切碎）
洋葱	半個（切碎）
椰菜	半個（切絲）

Ingredients

4 thick pork chops (marinated briefly
with salt and black pepper)
1 shallot (diced)
20 sprigs thyme
3 green apples (skin and core
removed; cut into eight)
1/2 cup white wine

Accompaniment

4 strips bacon (chopped)
1/2 onion (chopped)
1/2 cabbage (shredded)

做　法

1. 預熱焗爐 200°C/400°F。
2. 用一個可放入焗爐的鑊，用 1 湯匙油先爆香紅葱頭，將豬扒肥肉部分先煎香，再每邊煎約 2 分鐘至金黃色。
3. 將蘋果放入豬扒鑊中，即倒入白酒，加入百里香，把鑊放入焗爐內焗 20 分鐘。
4. 爆香煙肉、洋葱，放入椰菜兜勻，略煮 5 分鐘即可，取出待用。
5. 取少許伴碟放碟中，豬扒焗好後放上即成。

Method

1. Preheat the oven to 200°C/400°F.
2. Use a pan which can be put into oven to fry shallots in 1 tbsp of oil till fragrant. Place the fat parts of pork chops down to pan fry first. Fry each side for 2 minutes till golden brown.
3. Put the apples in the pan, pour white wine in, add the thyme. Put it in the oven to bake for 20 minutes.
4. Fry the bacon and onion till fragrant, add the cabbage to stir well. Cook for 5 minutes. Remove for use later.
5. Place a little accompaniment on a plate, arrange the pork chop on it to serve.

宜採用新鮮的百里香，味道更濃郁！
It is best to use fresh thyme for its added aroma!

Baked Ham

焗 火 腿

我試過在一個大 party 內，炮製了這個焗火腿，很受朋友的歡迎。這火腿的做法簡單、省時省力又美味，下次開 party 不妨考慮這個既體面、又可口的菜式！

我以又甜又酸的芒果醬作焗火腿的醬汁，芒果香與煙燻味絕對相配，令火腿更美味。

I once served a leg of ham at a big party which proved to be hugely popular amongst friends! This recipe is easy and effortless.

I use the sweet and sour mango sauce for baking ham. Mango blends well with the smoked taste, which adds more flavor to the ham.

材料

熟煙火腿	3.6-4.5 千克
啤酒	1 瓶（340ml）

芒果醬

芒果	2 個（切粒）
糖	1/4 杯
檸檬汁	1 湯匙

醬　料

芒果醬	2/3 杯
紅糖	1/3 杯
芥辣	1/4 杯
芥辣粉	1/4 湯匙
蘋果醋	1 湯匙
薑茸	2 茶匙

Ingredients

3.6 to 4.5 kg smoked ham
1 bottle beer (340 ml)

Mango Sauce

2 mangoes (cubed)
1/4 cup sugar
1 tbsp lemon juice

Sauce

2/3 cup mango sauce
1/3 cup brown sugar
1/4 cup mustard
1/4 tbsp mustard powder
1 tbsp apple cider vinegar
2 tsps minced ginger

✳ Remarks ✳

- 在火腿面剝格子紋，可令芒果醬更易滲入火腿內、更入味。
- 剩下的芒果醬可作芝士餅的餅面。
- 芒果醬的做法，可用於藍莓醬、草莓醬等。
- Making some cuts on the ham allows the mango sauce to easily penetrate into the meat.
- The residual mango sauce can be used as the topping for cheesecake.
- You may use the recipe for mango sauce to make blueberry sauce and strawberry sauce.

做　法

1. 預熱焗爐 160°C/320°F。
2. 將全部芒果醬料用慢火煮，煮至芒果軟爛即可，試味，如太酸，要下多點糖。
3. 把醬料混合。
4. 把啤酒倒在火腿上，蓋上錫紙，焗 1.5 小時。（期間要翻轉火腿一次）
5. 把醬料塗在火腿上，焗 15 分鐘，取出塗醬料，再焗 15 分鐘即可。

Method

1. Preheat the oven to 160°C / 320°F.
2. Cook all the mango sauce ingredients over low heat till it is mashed. Taste it to add sugar if it is too sour.
3. Combine the sauce ingredients well.
4. Pour the beer on the ham. Cover it with aluminum foil to bake for 1 1/2 hours (turn it once during the cooking time).
5. Rub the sauce on the ham and bake for 15 minutes. Remove it to spread sauce on it and further bake for 15 minutes.

焗盤的溫度有點高，但 Gigi 姐不怕燙手。
Baking tray can be very hot, but Gigi is undaunted!

Braised Pork Ribs

數字排骨

材　料

肉排　　　900 克（切成 2-3 吋）

調味料

紹興酒　　1 湯匙
鎮江醋　　2 湯匙
生抽　　　3 湯匙
砂糖　　　4 湯匙
水　　　　1 杯

Ingredients

900 g pork ribs (cut into pieces of 2-3 inches)

Seasoning

1 tbsp Shaoxing wine
2 tbsps Zhenjiang vinegar
3 tbsps light soy sauce
4 tbsps sugar
1 cup water

做　法

1. 全部調味料倒入已放了肉排的鍋內，大火煲滾再轉小火，不冚蓋，至水乾為止（中途將肉排翻身一次）。

Method

1. Put all the seasoning into a pot with the pork ribs. Bring it to the boil over high heat. Lower the heat; remove the cover and cook till the water is reduced (turn over the ribs once during cooking).

✳ Remarks ✳

- 數字排骨是每一位初入廚的朋友都覺得易學、易做的菜式，它的好味秘訣在於燜時不冚蓋，燜至汁稠為止。
- 爐火火侯的大小，都要格外留意，火太大，水分乾了肉卻未腍。火太小，肉爛了但煲內仍水汪汪。
- 要煮到汁稠肉稔，需要練習多次，習慣了自家爐火才能做到十全十美！
- Most new cooks find the Braised Pork Ribs easy to master. The secret of its flavor lies in taking off the cover while braising till the sauce thickens.
- Pay special attention to the heat. If the flame is too high, the water is dried but the meat is not cooked through. Too low a flame however while softening the meat, its sauce is too watery.
- It takes a lot of practice to have the meat tenderized with a thickened sauce. Get used to your own stove to get it perfect.

Clear Beef Brisket Soup
老大清湯腩

材　料
牛腩　　1.2 千克（出水半小時）

湯　底
八角　　　2 粒
甘草　　　4 片
黑椒粒　　1 湯匙
香葉　　　2 片
冰糖　　　1 湯匙
紅椒　　　1 隻
草果　　　1 個
桂皮　　　1 片
鹽　　　　1 茶匙
薑　　　　4 片
水　　　　適量（要浸過牛腩面）

Ingredients

1.2 kg beef brisket (scalded for 30 minutes)

Soup base

2 star anise
4 pieces liquorice root
1 tbsp black peppercorns
2 pieces bay leaf
1 tbsp rock sugar
1 red chili
1 black cardamon (Cao Guo)
1 piece cassia bark
1 tsp salt
4 slices ginger
water (enough to cover the beef)

那天，突然想吃清湯腩，想起老大，
於是通過電話請教，馬上買料炮製，
加上粿條，就是一頓清新可口的粿條
清湯腩！
老大是梅偉基先生，亦是「偉哥車仔
麵」的老板。

One day, I was in the mood for a bowl
of clear beef brisket soup. I then called
up Mr. Mei Wei Ji, the owner of Wei Ge
Push Cart Noodle for his tasty recipe.

做　法

1. 把已出水的牛腩及湯底香料同煮三小時，取出牛腩，待涼後才切件，牛腩可保持原件不會爛。
2. 享用前用熱湯淋熱牛腩即可。

Method

1. Boil the scalded beef and soup base for 3 hours. Remove the beef; slice when cool, to prevent it from breaking up.
2. Before serving, pour hot soup over the beef.

✳ Remarks ✳

- 香料有增香減牛腩羶味的作用
- While enhancing its flavor, the spices are also used to reduce the strong odor of beef.

Double-steamed Gingko Nuts and Liquorice with Honey

鮮白果甘草燉蜂蜜

材　料
鮮白果　2 包約 160 克（用水略沖）
甘草　　3 錢（沖洗乾淨）
蜂蜜　　1 湯匙
沸水　　4 杯

Ingredients

2 packets (about 160 g) fresh gingko nuts (washed)
12 g liquorice (washed)
1 tbsp honey
4 cups boiling water

做　法
1. 將所有材料放入燉盅內，用紗紙或微波爐保鮮紙封口，燉一小時即成。

Method

1. Put all the ingredients into a double-steaming pot. Seal it with mulberry paper or cellophane wrap used for microwave. Double-steam for 1 hour before serving.

甘草
Liquorice

✳ Remarks ✳

· 童年在廣州長大，很清楚記得平時會帶備幾片甘草在身，咀嚼時增加唾液，解口渴。這甘草糖水可治咳嗽，痰多及有哮喘者最宜飲用。

· Growing up in Guangzhou, I remember munching on liquorice to quench my thirst. This sweet soup heals coughing and it is suitable for those having asthma and phlegm.

Double-steamed Dried Figs and White Fungus with Rock Sugar

無花果雪耳燉冰糖

材 料

無花果	5 粒（一開二）
雪耳	50 克（溫水浸軟、剪去硬蒂、撕成小朵）
南北杏	40 克
薑	2 片
冰糖	120 克
沸水	6 杯

Ingredients

5 dried figs (halved)

50 g white fungus (softened in warm water; stems discarded and torn into pieces)

40 g bitter and sweet almonds

2 slices ginger

120 g rock sugar

6 cups boiling water

做 法

1. 全部材料放入燉盅內，用紗紙或微波爐保鮮紙封口，隔水燉 4 小時即成；或用電鍋熬 3 小時。
2. 此糖水冷熱飲皆宜。

Method

1. Place all the ingredients into a double-steaming pot. Seal it with mulberry paper or cellophane wrap used for microwave. Double-steam over water for 4 hours. You can cook this sweet soup in slow cooker and cook for 3 hours.
2. This sweet soup is good drinking cold or warm.

善待自己！在不出門的日子，翻翻家裏的食物箱，不
難找到這些乾貨。
看看書，覆兩封電郵，糖水就燉好了！好好享受吧！

Treat yourself to this sweet soup! It's not difficult to find
the required ingredients in your refrigerator. So, relax
at home and enjoy this concoction which can be done in
a jiffy!

Double-steamed Chinese White Cabbage with White Fungus and Dried Mushrooms

雪耳冬菇燉大白菜

炮製此湯最大的享受，是揭盅的一刻，那怡人的清香，動人心弦！

The greatest enjoyment preparing for this dish is the moment when the lid is lifted – such heart-warming aroma wafted from the soup!

材　料

雪耳　　40 克（溫水浸軟、剪去硬蒂）
冬菇　　6 朵（溫水浸軟、去蒂）
大白菜　400 克（洗淨後用沸水燙過）
薑　　　2 片
南北杏　40 克
沸水　　6 杯

Ingredients

40 g white fungus (soaked in warm water till softened; removed stems)
6 dried shittake mushrooms (soaked in warm water till softened; stems removed)
400 g Chinese white cabbage (washed then scalded in boiling water)
2 slices ginger
40 g sweet and bitter almonds
6 cups boiling water

做　法

1. 將全部材料放入燉盅內，用紗紙或微波爐保鮮紙封口，隔水燉四小時，加 1 茶匙鹽調味即可享用。

Method

1. Put all the ingredients into a stewing pot. Seal the lid with mulberry paper or cellophane wrap used for microwave. Double-boil over water for 4 hours. Season with 1 tsp of salt before serving.

Conch Soup with Lily Bulbs and Pears

百合雪梨響螺湯

新鮮響螺貴得買不下手，其實急凍響螺也不賴，而且價錢
比較平。為所愛煲一鍋愛心靚湯，想想也心甜！何況此湯
有滋補養顏、清熱止咳的功用呢！

Fresh conch is astronomical in cost. The frozen version is as
good and cheaper. Boil a pot of conch soup for the loved ones,
it enhances good health and beauty, in addition to expelling
Heat and stopping coughs.

材　料

急凍響螺　　4 個（解凍後用鹽洗淨，汆水）
瘦肉　　　　450 克（與響螺一齊汆水）
雪梨　　　　4 個（一開二，去核）
乾百合　　　75 克（用溫水略浸）
陳皮　　　　1 個（用溫水浸軟）
水　　　　　12 杯

Ingredients

4 frozen conches (washed with salt after
defrosting; scalded)
450 g lean pork (scalded with the
conches)
4 Chinese pears (halved and core
removed)
75 g dried lily bulbs (soaked briefly in
warm water)
1 dried tangerine peel (soaked in warm
water)
12 cups water

做　法

1. 煮沸水後，放入所有材料，先用大
　 火滾 10 分鐘，改為中慢火煲 3 小
　 時即可。
2. 各人口味不同，試味後才下鹽。

Method

1. Bring the water to the boil. Place
all the ingredients in the pot. Boil
over high heat for 10 minutes, then
turn down to medium low heat and
further boil for 3 hours.
2. Season with salt accordingly.

Tomato and Fish Slice Soup

番茄魚片湯

這個魚湯加入薄荷葉，入口時除了番茄及魚香外，更有清新的薄荷香。

Adding mint leaves into this fish soup makes it more aromatic.

材　料

番茄	4 個（去皮、切粒）
鯇魚片	200 克（切雙飛）
薄荷葉	8 片（切絲）
薑	2 片
沸水	3 杯

Ingredients

4 tomatoes (skin removed; cubed)
200 g grass carp (cut into butterfly shape)
8 mint leaves (shredded)
2 slices ginger
3 cups boiling water

將魚片切雙飛
Butterfly the fish.

做　法

1. 用少許油爆香薑片，加入番茄粒爆炒，注入沸水，煮透，約 5 分鐘。
2. 用 1 茶匙鹽抓洗魚片兩分鐘，兩次，沖凍水，用乾布索乾水分後，再加胡椒粉、蛋白及 1 茶匙薯粉略醃。
3. 湯煮至大滾後，將魚片小心鋪在番茄湯面，蓋上鍋蓋，熄火，讓熱力迫熟魚片，可保魚片不碎。吃前下 1 茶匙鹽，灑薄荷葉即可享用。

Method

1. Fry ginger slices in a little oil till fragrant. Add tomato cubes to stir fry. Pour boiling water in and boil for about 5 minutes.
2. Rub the fish slices with 1 tsp of salt for 2 minutes. Do it twice. Rinse with cold water. Use a clean towel to completely wipe off the moisture. Then add pepper, egg white and 1 tsp of sweet potato starch to marinate briefly.
3. After it is boiled, gently arrange the fish slices on top of the tomato soup. Cover. Switch off the flame to allow the heat to cook the fish. This is to ensure the shape of fish slices remains intact. Before serving, add 1 tsp of salt and sprinkle the mint leaves.

將魚片鋪在番茄湯面，蓋上鍋蓋，熄火，魚片熟後，下薄荷葉。

Place the fish slices on the surface of the tomato soup. Cover and switch off the flame. When the fish is cooked, add the mint leaves.

Coconut Soup

椰 子 湯

材　料

椰子肉	1 個（倒去椰汁，起肉，切塊）
無花果	4 個（切半）
黑豆	1 杯（略浸，沖淨）
冬菇	4 朵（溫水浸軟，剪去硬蒂，切半）
黑棗	4 粒
南北杏	1 湯匙
水	10 杯

Ingredients

1 coconut (juice discarded, flesh scooped up and cut into wedges)
4 dried figs (halved)
1 cup black beans (soaked briefly and washed)
4 dried shittake mushrooms (softened in warm water, stems discarded and halved)
4 black dates
1 tbsp sweet and bitter almonds
10 cups water

做　法

1. 全部材料倒入煲內，猛火煲滾幾分鐘後收慢火再煲兩小時即可。
2. 隨個人口味下鹽調味。

Method

1. Put all the ingredients into a pot. Bring it to the boil over high heat for a few minutes. Reduce the heat and boil it for 2 hours.
2. Season with salt accordingly.

✳ Remarks ✳

- 只要材料配搭得宜，湯不一定要加肉同煲才可口。這湯味道清甜、不膩。
- 買椰子的秘訣，是挑選重手、搖晃時有水聲的。
- 雖然椰子水的味道清甜，但性濕，對體弱者不宜，故要倒掉，不要用來煲湯。
- With the right accompaniments, meat need not be added to enhance the soup's flavor. This soup is refreshingly sweet without tasting greasy.
- Choose coconuts which are heavy; listen to the water swishing within when shaken.
- While coconut juice is sweet, it is believed to be Damp in the Chinese diet. Hence, it is not added to the soup.

Guava Soup
番石榴湯

材　料
青番石榴　　5-6 個（切塊）
豬腱　　　　450 克（汆水）
薑　　　　　4 片
水　　　　　12 杯

Ingredients

5-6 green guavas (cut into wedges)
450 g pork shin (scalded)
4 slices ginger
12 cups water

做　法

1. 煮沸水後，放入全部材料，先用猛
   火煲滾，再用慢火煲 2 小時即可。

Method

1. Boil the water and put in all the
   ingredients. Boil over high heat,
   turn to low heat and boil for 2
   hours. Serve.

＊ Remarks ＊

這湯我用從台灣進口的番石榴
（芭樂）煲的，湯味清甜。

I use Taiwanese guava for this
soup. It is sweet and light.

Vegetable Cream Soup

西式素菜湯

材　料

牛油　　　2 湯匙
洋葱　　　1 個約 50 克（切碎）
西芹　　　4 條約 50 克（切碎）
紅甜椒　　1 個（切粒）
紅蘿蔔　　1 條（去皮切碎）
馬鈴薯　　2 個約 150 克（去皮切粒）
沸水　　　10 杯
淡奶　　　1 杯
番茜碎　　2 湯匙

Ingredients

2 tbsps butter
1 onion (about 50 g) (chopped)
4 stalks celery (about 50 g)(chopped)
1 red capsicum (diced)
1 carrot (skin removed and chopped)
2 potatoes (about 150 g) (skin removed and diced)
10 cups boiling water
1 cup evaporated milk
2 tbsps chopped parsley

這湯的材料，你可按個人喜好及時令果菜而變化，例如加入粟米或青瓜等。如果想有咀嚼口感，可以在倒入淡奶前加入一些罐頭鷹豆。

You may add whatever seasonal or favorite vegetables into vegetable cream soup for varieties, such as corn or cucumber. For a different texture, add some canned chickpeas.

做　法

1. 用牛油爆香洋葱、西芹、紅甜椒、紅蘿蔔、馬鈴薯粒，倒入沸水，改用慢火煲 1 小時，關火。

2. 待菜湯稍涼，將菜湯分三次倒入攪拌機內打碎，再倒回湯煲以慢火煲滾。

3. 加入淡奶、鹽 1 茶匙，最後灑上番茜碎裝飾即可。

Method

1. Fry the chopped onion, celery, red capsicum, carrot and potato in butter till fragrant; pour the hot water in, turn to low heat and boil for 1 hour. Switch off the flame.

2. Cool the soup slightly, then divide the soup into three portions to blend in a blender three times, return it to the pot. Bring to the boil over low heat.

3. Add the evaporated milk and 1 tsp of salt. Sprinkle the chopped parsley to serve.

Papaya and Fish Tail Soup
木瓜魚尾湯

木瓜魚尾湯既清甜又有益，既然雞湯、骨湯煲得多，何妨換個口味，讓家人有均衡飲食。

Papaya and Fish Tail Soup is both sweet and healthy. While chicken and bone soups are the norm, it is perhaps time to change to a different flavor for balancing the diet in the family.

將魚尾兩面煎香，
放入魚袋，就可以
煲湯了。

Pan fry both sides of
the fish, put it in a
fish bag for boiling
soup.

材　料

鱅魚尾	1 份（即大魚尾）
薑	1 片
木瓜	1 個（去皮、去籽、切塊）
無花果	2 粒
紅棗	4 粒（去核）
花生	半杯（用溫水略浸）
腰果	半杯（用溫水略浸）
陳皮	半個（浸軟後刮去瓤）
百合	半杯（用溫水略浸，煲湯後 1 小時才下）
水	16 杯

Ingredients

1 bighead carp's tail

1 sliced ginger

1 papaya (skin removed, seeded and cut into chunks)

2 dried figs

4 red dates (stones removed)

1/2 cup peanuts (soaked briefly in warm water)

1/2 cup cashew nuts (soaked briefly in warm water)

1/2 dried tangerine peel (softened in water then pith removed)

1/2 cup dried lily bulbs (soaked briefly in warm water; add in soup only after it is boiled for 1 hour)

16 cups water

做　法

1. 用 1 湯匙油爆香 1 片薑，把魚尾用慢火煎香，然後放入魚袋內。

2. 凍水和紅棗倒入鍋內，水滾後放入全部材料（百合除外），猛火煲 10 分鐘，改用慢火煲 2 小時即可隨個人口味下鹽。

Method

1. Fry 1 sliced ginger in 1 tbsp of oil till fragrant, pan fry the fish tail over low heat until both sides are slightly brown. Put it in a fish bag.

2. Put the red dates in water, bring it to the boil. Put all the ingredients in (except dried lily bulbs) and boil over high heat for 10 minutes. Turn it down to low heat and further boil for 2 hours. Add salt as per personal taste.

Easy New York Cheesecake

簡易紐約芝士餅

只要將芝士餅的主要材料捏拿得準，就
可以從這基本餅料中變化。我喜歡這個
芝士餅有蜂蜜的香甜味做點綴，而且放
在水上燉焗，令餅的水分不致被抽乾。
第二層以酸忌廉鋪面，兩種不同的忌廉
重疊，吃時有質感有層次，這款基本紐
約芝士餅我真是百吃不厭。
今次我在這芝士餅餅面鋪上藍莓醬，又
有另一番風味；如有「焗火腿」時剩下
的芒果醬，也可伴芝士餅享用。

Once we master the basic recipe for
making cheesecakes, some varieties can
evolve from it. I like this cheesecake
because it has the sweet aroma of honey.
As it is steamed over water, it retains its
moisture. The second layer is spread with
sour cream. These two overlapping and
different creams give it a richer texture.
I usually spread blueberry sauce over the
cheesecake to create a different flavor. Any
left over mango sauce from Baked Ham
can be savored with the cake as well.

餅底材料

牛油溶液	2 湯匙
餅碎	1 杯

蛋糕料

忌廉芝士	500 克（室溫）
蜂蜜	5 湯匙（用微波爐「叮」一分鐘，軟化）
蛋	2 個
雲呢拿香油	1 茶匙
麵粉	2 湯匙（篩勻）
酸忌廉	500 克
糖	1/4 杯

藍莓醬

藍莓	2 杯
糖	1/4 杯
檸檬汁	2 湯匙

預備道具

9 吋甩底餅盤	一個
大焗盤	一個
蒸架	一個

Cake Base

2 tbsps melted butter
1 cup biscuit crumbs

Cake

500 g cream cheese
spread (room temperature)
5 tbsps honey (softened
over high heat in
microwave)
2 eggs
1 tsp vanilla essence
2 tbsps flour (sifted)
500 g sour cream
1/4 cup sugar

Blueberry Sauce

2 cups blueberry
1/4 cup sugar
2 tbsps lemon juice

Utensils

one 9 inch springform tin
one baking tray
one steamer rack

藍莓醬做法

將全部藍莓醬料用慢火煮，煮至藍莓軟爛即可，試味，如太酸，要下多點糖。

Method for Making Blueberry Sauce

Boil all the blueberries over low heat till they are mashed. Add sugar to taste.

芝士餅做法

1. 預熱焗爐 160°C/320°F。
2. 焗盤內放置蒸架，再注入半吋熱水（注意熱水不可以超過架子）。
3. 將餅底材料混合，壓在甩底餅盤底。
4. 用電動打蛋器的快速模式打勻忌廉芝士，改慢速，加入蜂蜜、蛋、麵粉及雲呢拿香油，充分攪勻後倒入餅底上，放入焗爐內蒸焗 45 分鐘，取出待涼。
5. 用電動打蛋器的快速模式打勻酸忌廉，逐少加入糖，打至不見糖為止。倒入「4」上輕輕掃勻，再焗 35 分鐘。
6. 待涼後，放入雪櫃數小時，鋪上藍莓醬就可享用了。

Method

1. Preheat the oven to 160°C/320°F.
2. Place the steamer rack on a baking tray, pour about 1/2 inch water in (watch that the hot water does not exceed the rack).
3. Combine well the ingredients for the cake base. Press hard onto the springform tin.
4. Use electric beaters to mix the cream cheese. Change to a slow speed, add the honey, eggs, flour and vanilla essence. Combine them well then pour on top of the cake base. Bake in the oven for 45 minutes. Remove to cool.
5. Beat the sour cream well, then add sugar gradually and beat until the sugar is dissolved. Pour into "4", gently smooth the surface. Bake for another 35 minutes.
6. Leave it to cool. Place in refrigerator for a few hours. Spread the blueberry sauce over to serve.

Chocolate Muffins

朱 古 力 鬆 餅

在北美洲，有已經調好的鬆餅粉出售，只需加奶，拌勻就可放入焗爐，人人都可做到鬆餅了，但味道可能會不盡你的口味。

我這個朱古力鬆餅雖然要自己量材料，但也不難做嘛，試試吧！

There are ready made muffin mix on sale in North America. All you need do is to add milk before baking it in an oven. However, while it is easy to make, the taste may not be what you prefer.

This chocolate muffin recipe is easy to make. Try it!

材　料

A

麵粉	2 杯	
發粉	1 茶匙	篩勻
鹽	1 撮	
朱古力粒	1 杯	

B

蛋	2 個
糖	半杯
朱古力奶	1 杯
菜油	2/3 杯

Ingredients

A

2 cups flour	
1 tsp baking powder	sifted
salt	
1 cup chocolate chips	

B

2 eggs
1/2 cup sugar
1 cup chocolate milk
2/3 cup vegetable oil

做　法

1. 預熱焗爐 175°C/350°F。
2. 篩勻 A 料後加入朱古力粒。
3. 順序打勻 B 料。
4. 把「2」倒入「3」內，攪拌均勻後倒入紙杯內，焗 20 分鐘即可。

Method

1. Preheat the oven to 175°C/350°F.
2. Sift Ingredients A, then add the chocolate chips.
3. Beat well Ingredients B in its order.
4. Pour "2" into "3", combine well then pour into paper cups. Bake for 20 minutes to serve.

Coffee Cake

咖啡蛋糕

材料

A
牛油	200 克（室溫）
糖	1 杯
雞蛋	4 個（室溫）

B
麵粉	2 杯
朱古力粉	6 湯匙
發粉	1 茶匙
鹽	1 撮

C
即溶咖啡粉	4 湯匙
咖啡酒	2 湯匙

* 拌勻

Ingredients

A

200 g butter (room temperature)
1 cup sugar
4 eggs (room temperature)

B

2 cups flour
6 tbsps cocoa powder
1 tsp baking powder
salt

C

4 tbsps instant coffee powder
2 tbsps Kahlua
* Mixed well

做 法

1. 預熱焗爐 175°C/350°F。
2. 先將 A 料的牛油打至奶白色，糖分 2 次加入，打至糖溶為止，逐個雞蛋加入，打至雞蛋與牛油糖料融合。
3. 將 B 料篩勻後倒入「2」內，慢速打勻。
4. 將 C 料倒入「3」內，以刮刀由底至面拌勻，倒入蛋糕模內，焗 30 分鐘即可。

Method

1. Preheat the oven to 175°C /350°F.
2. Beat the butter in Ingredients A till creamy, add sugar in two batches till it is dissolved. Add the eggs individually. Beat them till the eggs and butter are combined well.
3. Sift Ingredients B, then pour into "2", beat in slow speed.
4. Pour Ingredients C in "3". Use a spatula to stir from the bottom to the surface. Pour into a cake tin and bake for 30 minutes to serve.

咖啡香飄溢在空氣中，感覺好悠閒、好溫暖，
但一旦入喉，就覺很苦，製成蛋糕卻又是別有一番風味！

Coffee is aromatic but the taste is bitter.
However, use it in the cake, it is distinctly flavorful.

Cupcakes
小蛋糕

材　料

A
室溫牛油　1 杯
糖　　　　3/4 杯
* 打成奶白色

B
麵粉　　　2 杯
鹽　　　　1 撮
發粉　　　1 茶匙
* 篩勻

C
蛋　　　　5 個
乾椰茸　　半杯（用 175℃/350℉ 焗
　　　　　2 分鐘至金黃色）
椰子香油　數滴

Ingredients

A
1 cup butter
(room temperature)
3/4 cup sugar
* beat into creamy color

B
2 cups flour
salt
1 tsp baking powder
* sifted

C
5 eggs
1/2 cup desiccated coconut (baked in
oven at 175℃/350℉ for 2 minutes to
golden brown)
a few drops coconut essence

小蛋糕自有她的地位與吸引力，讓人忍
不住一口一個的吞下肚，就因為它的嬌
小，也不怕吃過量了。
炮製小蛋糕一點也不難，但用匙羹舀糊
料進細細的蛋糕模則有點困難，稍一不
慎會將糊料掛在模邊或滴在工作檯上，
所以我用擠花袋將糊料擠在模裏，既方
便又快捷。

The petit cupcakes are adorable
and given its size, there is no fear of
overindulging in it.
It is simple to make cupcakes. However,
it is a little tricky to spoon the mixture
into tiny cake moulds. I have opted
to use a pastry bag to fill the mould –
convenient and neat!

做　法

1. 預熱焗爐 175˚C/350˚F。
2. 打勻材料 A 後，逐一加入蛋，打勻（分 5 次加入蛋，蛋與牛油糖混合料打至融合後才加入另一個蛋）。
3. 分三次把 B 料加入「2」，慢慢攪拌，再加入椰茸及椰子香油，把糊料擠入小蛋糕模中，焗 20-25 分鐘即成。

Method

1. Preheat the oven till 175˚C/350˚F.
2. Beat Ingredients A well, gradually add the eggs and beat well (add the eggs individually 5 times. Beat the egg and butter sugar mixture till blended before adding another egg).
3. By three times, add Ingredients B to "2", stir slowly. Add the desiccated coconut and coconut essence. Pour the mixture into small cake moulds and bake for 20-25 minutes. Serve.

✳ Remarks ✳

- 要逐個雞蛋與牛油糖料打勻，不要一下子倒入所有雞蛋，令牛油糖料不能與雞蛋融合會呈「豆腐渣」狀，焗後的蛋糕會很「實」。
- Beat the eggs individually with the butter and sugar mixture. Do not add the eggs all at once, or it will not blend well and the mixture will curdle. The cake as a result will be hardened.

Banana Cake

香蕉蛋糕

香蕉蛋糕的食譜，收集不下五六個，
唯有這個做法最鬆軟。
主要原因是香蕉於急凍後其纖維斷層，
焗好後就不會因香蕉太重而沉墜；
另外，檸汁的作用是讓梳打粉藉酸性發揮升起的效果，
蛋糕就會鬆軟而濕潤。
我用菜油代替牛油，
較為健康、輕怡，食得放心！

Out of the many recipes on banana cake I collected,
this one proves to make the fluffiest version.
The main reason is that freezing breaks the fibres
which makes it lighter in baking.
Moreover, lemon juice enhances the baking soda's rising intensity
to make the cake fluffy and moist.
I replace butter with the lighter and healthier vegetable oil.

材　料

熟香蕉	3 隻（置冰格雪至斷纖維，用前解凍）
牛奶	1 杯（與香蕉打勻，待用）
麵粉	2 杯
梳打粉	1 茶匙
發粉	1.5 茶匙
糖	3/4 杯
菜油	3/4 杯
蛋	2 個（打勻）
檸檬汁	半個份量

麵粉、梳打粉、發粉 ── 三樣粉同篩勻

Ingredients

3 ripe bananas (freeze till the fibres are broken; defrost before use)
1 cup milk (blended well with bananas)
2 cups flour
1 tsp baking soda
1 1/2 tsps baking powder
} sifted well
3/4 cup sugar
3/4 cup vegetable oil
2 eggs (beaten well)
juice of 1/2 lemon

做　法

1. 預熱焗爐 175°C/350°F。
2. 熟香蕉、牛奶放入大碗內，打勻。
3. 倒入已篩過的麵粉，用打蛋器慢速打 1 分鐘，再逐樣倒入其餘材料，拌勻。將蛋糕糊倒入模內，焗 45 分鐘即可。

Method

1. Preheat the oven to 175°C /350°F.
2. Put the ripe bananas and milk into a big bowl and mix well.
3. Pour into the sifted flour. Beat in slow speed with the electric beaters for 1 minute, gradually pour the rest of the ingredients in, stir well. Pour the batter into cake moulds. Bake for 45 minutes to serve.

Honey and Lemon Cheesecake
檸 蜜 芝 士 餅

芝士餅是許多女士不能抗拒的誘惑，
即使是 on diet 也忍不住啖一口。

Cheesecake is so irresistible even to those who are on diet!

蛋糕料

忌廉芝士	500 克（室溫）
蜂蜜	半杯
檸檬	1 個（將皮刮出留用，榨汁約 1/4 杯，待用）
麵粉	3 湯匙（篩過）
雞蛋	2 個（打勻）
蛋黃	2 個（加入雞蛋中同拌勻）

餅　底

餅碎	1 杯
牛油溶液	2 湯匙

餅　面

打發忌廉

Cake

500 g cream cheese spread (room temperature)

1/2 cup honey

1 lemon (grated the zest for later use; extract juice about 1/4 cup)

3 tbsps flour (sifted)

2 eggs (beaten)

2 egg yolks (combined with the eggs)

Cake Base

1 cup biscuit crumbs

2 tbsps melted butter

Topping

whipped cream

做　法

1. 預熱焗爐 150°C/300°F，焗盤中加蒸架，
   倒入熱水，約一吋高。
2. 把餅碎與牛油混合，平鋪在甩底盤中，壓
   緊，待用。
3. 將忌廉芝士置於大碗中，以打蛋機打勻，
   順序加入材料，每樣材料打至混合為止，
   才加第二樣，攪拌至滑，倒入餅底中。
4. 將芝士餅盤置蒸架上，蒸焗 1 小時至面呈
   微黃色即可取出，待涼。
5. 放入雪櫃 3 小時，塗上忌廉，即可享用。

Method

1. Preheat the oven to 150°C/300°F. Place a
   steamer rack on the tray, pour hot water on
   it (about 1 inch high).
2. Combine the biscuit crumbs and butter,
   spread it on a springform tin. Press down
   hard. Reserve for use.
3. Put the cream cheese into a big bowl. Beat
   well with egg beaters. Add the ingredients
   in its order, mix each ingredient in well
   before adding another one. Combine till
   smooth. Pour into the cake tin.
4. Place the tin on the steamer rack. Bake for
   1 hour till the surface turns slightly golden
   brown. Cool.
5. Place it in the refrigerator for 3 hours.
   Spread the top of cheesecake with
   whipped cream. Serve.

芝士餅焗後，放在焗爐待至冷卻，
才不會如圖示般裂開。

To avoid cracking of the surface, leave the
baked cheesecake in the oven for cooling.

烹飪充滿樂趣。在拍攝當天，我將從電視看來的用膠樽吸蛋黃法示範給工作人員，看得他們大樂。

Cooking is fun. On the day of photography, I demonstrated using a plastic bottle to suck up egg yolks for the crew. How elated they are!

Matcha Panna Cotta
抹茶奶凍

在北美洲很容易就買到做甜品的材料，
雖然價錢也不便宜（與店內的現成製成品相約），
但怎樣也不及香港的貴；所以，許多時我們都情願買現成的。
不過，近年發現有些食物達不到安全標準，還是自己動手吧！

In North America, the cost of buying ingredients for making desserts
is just as expensive as the readymade products.
However, they are still cheaper than buying them in Hong Kong.
As such, most times we resort to readymade ones.
So, why not try making them ourselves?
At least the food safety standard can be guaranteed!

材　　料

魚膠片	4 片（冷水浸 5 分鐘）
牛奶	1 杯
淡奶	3 杯
日本抹茶粉	2 茶匙
蜂蜜	1/3 杯
雲呢拿香油	2 茶匙

Ingredients

4 gelatin sheets
(soaked in cold water for 5 minutes)
1 cup milk
3 cups evaporated milk
2 tsps Japanese green tea powder
1/3 cup honey
2 tsps vanilla essence

做　　法

1. 牛奶及 1 杯淡奶用中火加熱，加入抹茶粉及魚膠片，邊加邊拌至溶化後離火。
2. 離火後加入蜂蜜、其餘 2 杯淡奶及雲呢拿香油，拌勻後即可倒入小杯內，冷藏 3 小時即可。

Method

1. Heat the milk and 1 cup of evaporated milk over medium heat, add the green tea powder and gelatin sheets, stirring it while adding till it is dissolved. Switch off flame.
2. Remove it from the flame, add the honey, the remaining 2 cups of evaporated milk and vanilla essence, mix well then pour into small cups. Refrigerate for 3 hours before serving.

Baked "Nian Gao"

焗年糕

這是我最早學做的年糕，不知何故，食譜無故失蹤，直至拍攝前整理文件，食譜重現眼前，喜出望外！
焗年糕，與傳統的蒸年糕口味截然不同，糕面脆脆的、又香口，別有一番風味。

I used this recipe when I first learnt how to make nian gao (rice cake). It was however misplaced for many years. I was ecstatic when it was found!

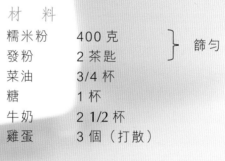

材　料
糯米粉	400 克	}　篩勻
發粉	2 茶匙	
菜油	3/4 杯	
糖	1 杯	
牛奶	2 1/2 杯	
雞蛋	3 個（打散）	

Ingredients

400 g glutinous rice flour } sifted
2 tsps baking powder
3/4 cup vegetable oil
1 cup sugar
2 1/2 cups milk
3 eggs (beaten)

做　法
1. 預熱焗爐 175°C/350°F。
2. 油、糖、牛奶齊攪勻,逐少加入雞蛋。
3. 把粉料倒入「2」中,用打蛋器慢速拌勻,
　 倒入已抹油的盆中,焗 45 分鐘即可。

Method
1. Preheat the oven to 175°C /350°F.
2. Combine the oil, sugar and milk well.
   Gradually add the eggs little by little.
3. Pour the flour into "2". Beat well with
   egg beaters in slow speed. Pour the
   mixture into a greased pan and bake for
   45 minutes to serve.

Tuiles aux Amandes
杏仁薄脆片

這是我在溫哥華烹飪學校跟馬師傅學的
甜品。

正式的杏仁薄脆應該做成瓦片狀，但道
具多了，步驟也多了，不合乎我這懶人
要求簡易的原則，反而將焗好的薄脆，
隨意掰開，更自然，更可愛！

可能你會有疑問：為甚麼要加入馬鈴薯
粉？它的作用是不會讓麵糊起筋，焗後
更香脆。

I learnt this recipe from Chef Ma at a
cooking school in Vancouver.

The proper Tuiles aux Amandes involves
many props and steps to make it.

However, to keep it simple and easy, I
randomly tear the baked Tuiles to make it
appear more natural and pleasing!

Why add potato flour? You may ask. It is
meant to lighten the dough. It is crispier
after baking.

材 料

蛋白	2 個
幼砂糖	1/3 杯
牛油	1/4 杯（室溫）
杏仁香油	1/3 茶匙
麵粉	1/4 杯
馬鈴薯粉	1/4 杯
杏仁片	1/3 杯（用 160℃/320℉ 焗 5 分鐘至微黃）

Ingredients

2 egg whites
1/3 cup caster sugar
1/4 cup butter (room temperature)
1/3 tsp almond essence
1/4 cup flour
1/4 cup potato flour
1/3 cup almond flakes (bake in 160°C/
320°F for 5 minutes till slightly golden
brown)

做　法

1. 預熱焗爐 175℃/350℉。
2. 先把蛋白及糖拌勻，再順序加入其餘材料，拌勻後放入雪櫃 30 分鐘待用。
3. 烤盤鋪上牛油紙，取「2」抹在紙上，越薄越好，焗 6-8 分鐘至微黃即可取出。
4. 冷卻後隨意掰開，大小不拘，即可享用。

Method

1. Preheat the oven to 175℃/350℉.
2. Blend the egg whites and sugar well, gradually add the rest of the ingredients in its order. Mix well then place it in the refrigerator for 30 minutes for later use.
3. Lay a baking sheet on the baking tray. Spread "2" on it as thinly as possible. Bake for 6-8 minutes till slightly golden brown. Remove.
4. Leave it to cool and randomly tear them apart and serve.

✳ Remarks ✳

正如前述，我是個以簡單、方便為原則的人，一些烹飪小智慧，會令煮食更有趣、簡易。在鋪牛油紙之前，灑幾滴水，便可以固定牛油紙，在抹麵糊時不會「郁來郁去」；而且在烤盤上鋪牛油紙，取出食物時除方便外，更在清洗時省功夫。

Little cooking tips do make the process more fun and easier such as sprinkle a few drops of water on the tray before laying on a baking sheet in order to secure it. When spreading the dough on the baking sheet, the paper stays still. Laying baking sheet on baking trays makes removing the food more convenient and cleaning easier.

Savory Cheese Scones

芝士鹹鬆餅

邀請朋友回家 high tea 或訪友做手信，有這個美味的芝士鹹鬆餅，一定會讓朋友愛透你。

這鬆餅的做法比麵包簡易得多，做麵包要發酵兩次，對都市人來説會比較繁複、疲累。而做鬆餅是用手指尖的溫度捏溶牛油，拌勻全部材料只需用膠刮就成，方便、易做。而且吃剩的，可以放在冰格貯存（保鮮期約一星期），吃時再翻熱，同樣美味。

This is a fantastic gift for friends or you can serve it at high teas at home.

This recipe is easier to handle than bread-making. To the city folks, letting yeast rise twice seems a chore. This recipe however allows us to let the warmth of our finger tips melt the butter; a spatula is sufficient to mix well all ingredients. Simple and convenient. The balance scones can be left in the freezer for a week. Heat them up when eating, it will be just as delicious.

材料

A

麵粉	1 杯半
黃粟米粉	半杯
鹽	1 撮
發粉	2 茶匙
梳打粉	1 茶匙

* 篩勻

B

凍牛油粒	4 湯匙（112 克）
巴馬臣芝士碎	2 杯
葱花	半杯
牛油奶	1 杯
橄欖油	1/3 杯

Ingredients

A

1 1/2 cups flour
1/2 cup yellow cornmeal
salt
2 tsps baking powder
1 tsp baking soda
* sifted well

B

4 tbsps frozen cubed butter (112 g)
2 cups grated parmesan cheese
1/2 cup chopped spring onion
1 cup buttermilk
1/3 cup olive oil

做　法

1. 預熱焗爐 200°C/400°F。
2. 先把 A 料篩勻，加入牛油粒，用手指溫度捏溶牛油混合在其中。
3. 加入芝士碎、葱花，以膠刮拌勻，倒入牛油奶及橄欖油輕輕拌勻。
4. 把混合料分別舀進鬆餅模內，焗 20 分鐘至金黃，取出待涼才享用。
5. 此餅可以檸檬牛油伴食，將室溫牛油半杯混合 1 個份量的檸檬皮茸，打發至混合即可。

Method

1. Preheat the oven to 200°C/400°F.
2. Sift Ingredients A, add the cubed butter, use finger tips to melt the butter to mix well.
3. Add parmesan cheese and chopped spring onion, use a spatula to combine well. Pour buttermilk and olive oil to gently mix it.
4. Pour the mixture into muffin moulds. Bake for 20 minutes till golden brown. Cool before serving.
5. Savor it with lemon butter - whisk 1/2 cup room temperature butter with lemon zest from 1 lemon.

Orange Scones

香橙鬆餅

鬆餅，配上濃香的紅茶，別具風格。
除了甜味鬆餅外，亦可加入芝士、煙
肉等做成鹹鬆餅，你們可參考 P.158
的芝士鹹鬆餅。鬆餅好味的秘訣是切
拌時千萬不能拌太耐，否則會起筋，
做出來的鬆餅就不鬆軟了。

Scones are a great treat for the
English afternoon tea. Besides the
sweet versions, you may include
cheese and bacon to make savory ones.
Please refer to page 158 on the recipe
for Savory Cheese Scones. The secret
for good scones lies in not stirring too
much when mixing the dough, or the
scones will not be fluffy.

材　料

麵粉	3 杯	} 篩勻
發粉	1 1/2 湯匙	
糖霜	1 杯	
橙皮茸	1 個份量	
牛油	112 克	
	（切粒，紅棗般大小）	
蛋黃	2 個（打散）	
小紅莓乾	2/3 杯（用熱水浸至軟身）	

塗面蛋液

| 蛋 | 1 個 |
| 水 | 2 茶匙 |

* 拌勻，用篩過濾，隔去硬塊

Ingredients

3 cups flour
1 1/2 tbsps baking powder } sifted well
1 cup icing sugar
zest of 1 orange
112 g butter (diced into the size of red
date)
2 egg yolks (beaten)
2/3 cup dried cranberries (softened in
hot water)

Egg Wash

1 egg
2 tsps water
* combined well and sifted

做　法

1. 預熱焗爐 190°C/375°F。
2. 麵粉、發粉篩勻後，置大盆內，加入糖霜及橙皮茸，拌勻後再加入其餘材料，用膠刮切拌幾下即可（拌太久會起筋）。
3. 桌面灑麵粉，把「2」倒出略切拌，再擀成一吋半厚的麵糰，用模型印出圓形，掃上蛋液，放入焗爐焗 12-18 分鐘即可。

Method

1. Preheat the oven to 190°C/375°F.
2. Sift the flour and baking powder in a big bowl. Add the icing sugar and orange zest. Mix well then add the rest of the ingredients. Lightly knead the dough with a spatula only a few times (too many will harden the scones).
3. Sprinkle some flour on the counter top, lightly cut and fold the dough. Roll out the dough into 1 1/2 inches in thick. Cut with mould to a round shape. Glaze the egg wash on top. Bake in the oven for 12-18 minutes to serve.

謝謝您們，令這書內容更豐富

A Big Thank You to all for enriching the Book!

- Danny Kwong~奶油龍蝦、簡易紐約芝士餅、檸蜜芝士餅
  Danny Kwong for Butter Lobster, Easy New York Cheesecake and Honey and Lemon Cheesecake

- Lisa Ng~Lisa豬扒
  Lisa Ng for Lisa's Pork Chops

- 楊煥素~黑椒雞髀
  Sue Wan for Stewed Black Pepper Chicken Drumsticks

- 梅偉基~老大清湯腩
  Mei Wei Ji for Clear Beef Brisket Soup

- 馬師傅~杏仁薄脆片
  Master Ma for Tuiles aux Amandes

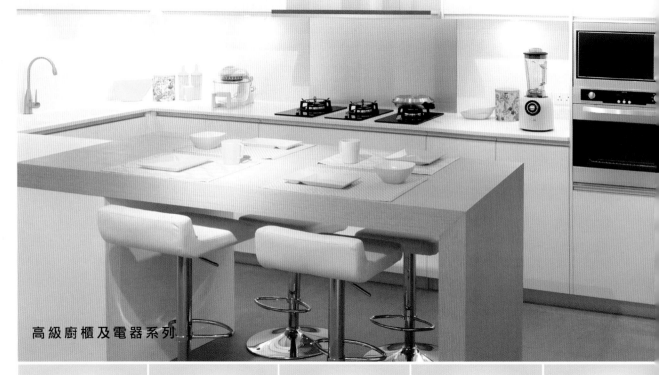

電器 │ 廚櫃

30 開心相伴30載

創新 · 多功能 · 節能

貼心設計　健康生活

光波萬能煮食鍋

自然養生機

8合1電子炸鍋

高速多層電蒸鍋

自家製雪糕機

自家製麵包機

智能電炒鍋

保健養生鍋

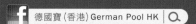

濃情美味
為最愛的人下廚

Favorite Food
for the Beloved

作者 / Author
黃淑儀 / Gigi Wong

策劃/編輯 / Project Editor
Catherine Tam

翻譯 / Translator
Patricia Mok

攝影 / Photographer
Imagine Union

美術統籌及設計 / Art Direction & Design
Amelia Loh

出版者 / Publisher
Forms Kitchen
香港鰂魚涌英皇道1065號 / Room 1305, Eastern Centre, 1065 King's Road,
東達中心1305室 / Quarry Bay, Hong Kong
電話 / Tel: 2564 7511
傳真 / Fax: 2565 5539
電郵 / Email: info@wanlibk.com
網址 / Web Site: http://www.formspub.com
http://www.facebook.com/formspub

瀏覽網站　會員申請

發行者 / Distributor
香港聯合書刊物流有限公司 / SUP Publishing Logistics (HK) Ltd.
香港新界大埔汀麗路36號 / 3/F., C&C Building, 36 Ting Lai Road,
中華商務印刷大廈3字樓 / Tai Po, N.T., Hong Kong
電話 / Tel: 2150 2100
傳真 / Fax: 2407 3062
電郵 / Email: info@suplogistics.com.hk

承印者 / Printer
中華商務彩色印刷有限公司 / C & C Offset Printing Co., Ltd.

出版日期 / Publishing Date
二〇一三年七月第一次印刷 / First print in July 2013
二〇一七年一月第五次印刷 / Fifth print in January 2017